KU-226-926

# IMPLEMENTING VOICE OVER IP

# IMPLEMENTING VOICE OVER IP

**BHUMIP KHASNABISH**
Lexington, Massachusetts, USA

A JOHN WILEY & SONS, INC. PUBLICATION

Copyright © 2003 by John Wiley & Sons, Inc. All rights reserved.

Published by John Wiley & Sons, Inc., Hoboken, New Jersey.
Published simultaneously in Canada.

No part of this publication may be reproduced, stored in a retrieval system, or transmitted in any form or by any means, electronic, mechanical, photocopying, recording, scanning, or otherwise, except as permitted under Section 107 or 108 of the 1976 United States Copyright Act, without either the prior written permission of the Publisher, or authorization through payment of the appropriate per-copy fee to the Copyright Clearance Center, Inc., 222 Rosewood Drive, Danvers, MA 01923, 978-750-8400, fax 978-750-4470, or on the web at www.copyright.com. Requests to the Publisher for permission should be addressed to the Permissions Department, John Wiley & Sons, Inc., 111 River Street, Hoboken, NJ 07030, (201) 748-6011, fax (201) 748-6008, e-mail: permreq@wiley.com.

Limit of Liability/Disclaimer of Warranty: While the publisher and author have used their best efforts in preparing this book, they make no representations or warranties with respect to the accuracy or completeness of the contents of this book and specifically disclaim any implied warranties of merchantability or fitness for a particular purpose. No warranty may be created or extended by sales representatives or written sales materials. The advice and strategies contained herein may not be suitable for your situation. You should consult with a professional where appropriate. Neither the publisher nor author shall be liable for any loss of profit or any other commercial damages, including but not limited to special, incidental, consequential, or other damages.

For general information on our other products and services please contact our Customer Care Department within the U.S. at 877-762-2974, outside the U.S. at 317-572-3993 or fax 317-572-4002.

Wiley also publishes its books in a variety of electronic formats. Some content that appears in print, however, may not be available in electronic format.

*Library of Congress Cataloging-in-Publication Data is available.*

ISBN 0-471-21666-6

Printed in the United States of America

10 9 8 7 6 5 4 3 2 1

*This book is dedicated to:*

*Srijesa, Inrava, and Ashmita;*
*My parents, sisters, and brothers;*
*My teachers, present and past*
*Colleagues, and friends; and*
*All of those who consciously and*
*honestly contributed to*
*making me what I am today.*

# CONTENTS

# PREFACE

In general, voice transmission over the Internet protocol (IP), or VoIP, means transmission of real-time voice signals and associated call control information over an IP-based (public or private) network. The term *IP telephony* is commonly used to specify delivery of a superset of the advanced public switched telephone network (PSTN) services using IP phones and IP-based access, transport, and control networks. These networks can be either logically overlayed on the public Internet or connected to the Internet via one or more gateways or edge routers with appropriate service protection functions embedded in them. In this book, I use *VoIP* and *IP telephony* synonymously, most of the times.

This book grew out of my participation in many VoIP-related projects over the past several years. Some of the early projects were exploratory in nature; oscillators had to be used to generate certain tones or signals, and oscilloscopes were used to measure the dial-tone delay, call setup time, and voice transmission delay. However, as the technology matured, a handful of test and measurement devices became available. Consequently, we turned out to be better equipped to make more informed decisions regarding the computing and networking infrastructures that are required to implement the VoIP service. Many of the recent VoIP-related projects in the enterprise and public network industries involve specifying a VoIP service design or upgrading an existing VoIP service platform to satisfy the growth and/or additional feature requirements of the customers. These are living proof of the facts that all-distance voice transmission service providers (retailers and wholesalers) and enterprise network designers are seriously deploying or considering the deployment of VoIP services in their networks.

This book discusses various VoIP-related call control, signaling, and transmission technologies including architectures, devices, protocols, and service requirements. A testbed and the necessary test scripts to evaluate the VoIP service and the devices are also included. These provide the essential knowledge and tools required for successful implementation of the VoIP service in both service providers' networks and enterprise networks. I have organized this information into nine chapters and three appendixes.

Chapter 1 provides some background and preliminary information on introducing the VoIP service for both residential and enterprise customers. I also discuss the evolution of the monolithic PSTN switching and networking infrastructures to more modular, distributed, and open-interface-based architectures. These help rapid rollout of value-added services very quickly and cost-effectively.

Chapter 2 reviews the emerging protocols, hardware, and related standards that can be used to implement the VoIP service. These include the bandwidth efficient voice coding algorithms, advanced packet queueing, routing, and quality of service delivery mechanisms, intelligent network design and dimensioning techniques, and others.

No service can be maintained and managed without proper signaling and control information, and VoIP is no exception. The problems become more challenging when one attempts to deliver real-time services over a routed packet-based network. Chapter 3 discusses the VoIP signaling and call control protocols designed to provide PSTN-like call setup, performance, and availability of services.

Next, I discuss the criteria for evaluating the VoIP service. In traditional PSTN networks, the greater the end-to-end delay, the more significant or audible becomes the return path and talker echo, resulting in unintelligible speech quality. Therefore, hardware-based echo cancellers have been developed and are commonly used in PSTN switches to improve voice quality. In packet networks, in addition to delay, packet loss and variation of delay (or delay jitter) are common impairments. These impairments cause degradation of voice quality and must be taken into consideration when designing an IP-based network for delivering the VoIP service. I discuss these and related issues in Chapter 4.

Various computing and networking elements of a recently developed VoIP testbed are considered in Chapter 5. This testbed has been used both to prototype and develop operational engineering rules to deliver high-quality VoIP service over an IP network.

Chapters 6, 7, and 8 focus on various possible VoIP deployment scenarios in enterprise networks, public networks, and global enterprises. Enterprise networks can utilize VoIP technology to offer voice communications services both within and between corporate sites, irrespective of whether these sites are within the national boundary or anywhere in the world. In the public networking arena, the VoIP service can be introduced in PSTN, cable TV, and wireless local loop–based networks for local, long-distance, and international calls.

Chapter 9 is the final chapter. In addition to presenting some concluding remarks and future research topics, I provide some guidelines for implementing the VoIP service in any operational IP network. These include the reference architectures, implementation agreements, and recommendations for network design and operations from a handful of telecom, datacom, and cable TV network/system standardization organizations.

Implementation of a few techniques that can be utilized to measure the call set performance and bulk call-handling performance of the VoIP network elements (e.g., IP-PSTN gateways, the VoIP call server) are presented in Appendixes A and B. Appendix C illustrates experimental evaluation of the quality of transmission of voice signal and DTMF digits in both PSTN-like and IP networks with added packet delay, delay jitter, and packet loss scenarios.

In the Glossary of Acronyms and Terms, definitions and explanations of widely used VoIP terms and abbreviations are presented.

Finally, I hope that you will enjoy reading this book, and find its contents useful for your VoIP implementation projects. As the technologies mature or change, much of the information presented in this book will need to be updated. I look forward to your comments and suggestions so that I can incorporate them in the next edition of this book. In addition, I welcome your constructive criticisms and remarks. My e-mail addresses are b.khasnabish@ ieee.org and bhumip@acm.org (www1.acm.org/~bhumip).

BHUMIP KHASNABISH

Battle Green
Lexington, Massachusetts, USA

# ACKNOWLEDGMENTS

My hat goes off to my children who inspired me to write this book. They naively interpreted the VoIP network elements as the legos during their visits with me to many of the VoIP Labs. This elucidation is more realistic when one considers the flexibility of the VoIP network elements to help rapid rollout of new and advanced services.

By posing the issues from many different viewpoints, my friends and colleagues from GTE Laboratories (now a part of Verizon) and Verizon Laboratories helped me understand many of the emerging VoIP related matters. Accordingly, my special thanks are due to—among others—Esi Arshadnia, Nabeel Cocker, Gary Crosbie, John DeLawder, Elliot Eichen, Ron Ferrazzani, Bill Goodman, Kathie Jarosinski, Naseem Khan, Alex Laparidis, Steve Leiden, Harry Mussman, Winston Pao, Edd Rauba, Gary Trotter, and George Yum. I have touched on several topics in this book, and many of them may need further investigations for network evolution.

This book would not have been possible without the support and encouragements I received from Dean Casey, Roger Nucho, Prodip Sen, and Mike Weintraub of Verizon Laboratories. I am really indebted to them for the challenging environment they provide here in the Labs.

The acquisition, editorial, and production staff of the Scientific, Technical, and Medical (STM) division of the John Wiley and Sons, Inc. deserves recognition for their extraordinary patience and perseverance. My special thanks go to Brendan Codey, Philip Meyler (who helped me at the initial phases of this project), Val Moliere (Editor), Andrew Prince, Christine Punzo, Kirsten Rohstedt, and George Telecki.

Finally, my wife, children, and relatives at home and abroad spared me of many duties and responsibilities when I was preparing the manuscript for this book. Their patience, heartfelt kindness, and sincere forgiveness cannot be expressed in words. I am not only grateful to them, but also earnestly hope that they will undertake such endeavors sometime in the future.

BHUMIP KHASNABISH

Barnstable Harbor
Cape Cod, Massachusetts, USA

# 1

# BACKGROUND AND INTRODUCTION[1]

Implementation of real-time telephone-quality voice[2] transmission using the Internet protocol (IP, the Internet Engineering Task Force's [IETF's] request for comment [RFC] 2460 and RFC 791) is no longer as challenging a task as it was a few years ago [1,2]. In this introductory chapter, I define the instances and interfaces of both public switched telephone networks (PSTN) and corporate or enterprise communication networks where voice over IP (VoIP) can be implemented. The goals of VoIP implementation are to achieve (a) significant savings in network maintenance and operations costs and (b) rapid rollout of new services. The objective is to utilize open, flexible, and distributed implementation of PSTN-type services using IP-based signaling, routing, protocol, and interface technologies. To achieve this, it is necessary to change the mindset of those responsible for the design and operations of traditional voice services networks. Furthermore, one has to be ready to face the challenging problems of achieving reliability, availability, quality of service (QoS), and security up to the levels that are equivalent to those of the PSTN networks.

I discuss two paradigms for implementing the VoIP service in the next section, and then present a few scenarios in which VoIP-based telephone service can be achieved for both residential and enterprise customers. A functionally layered architecture is then presented that can be utilized to facilitate the separation of call control, media adaptation, and applications and feature hosts. Finally, I describe the organization of the rest of the book.

---

[1] The ideas and viewpoints presented here belong solely to Bhumip Khasnabish, Massachusetts, USA.

[2] 300 to 3400 Hz (or 3.4 KHz) of analog speech signal.

## THE PARADIGMS

The following two paradigms are most prevalent for implementation of the VoIP service:

- Server, router, and personal computer (PC)/plain old telephone service (POTS) phone-based (mostly) flat network and
- PSTN switch and mainframe computers, VoIP gateway (GW)[3] and gate-keeper (GK),[4] SS7 signaling gateway (SG),[5] and the POTS-phone/PC-based (mostly) hierarchical network.

In order to provide VoIP and IP telephony services, PCs need to be equipped with a full-duplex audio or sound card, a modem or network interface card (NIC) such as an Ethernet[6] card, a stereo speaker, a microphone, and a software package for telephone (keypad, display, feature buttons, etc.) emulation. Hardware-based IP phones can be used with a traditional PSTN network using special adapter cards—to convert the IP packets into appropriate TDM-formatted voice signals and call control messages—as well.

In the server-router-based networking paradigm, the servers are used for hosting telephony applications and services, and call routing is provided by traditional packet routing mechanisms. In the other case, the telephone features and services can still reside in the PSTN switch and/or the adjacent mainframe computer, and the packet-based network elements—for example, the VoIP GW, GK, and SG—can offer a sufficient amount of signaling, control, and transport mediation services. Call routing in this case follows mainly the traditional hierarchical call routing architecture commonly utilized in the PSTN networks.

The details of network evolution and service, network, control, and management architectures depend on the existing infrastructures and on technical, strategic, and budgetary constrains.

## VoIP FOR RESIDENTIAL CUSTOMERS

In the traditional PSTN networks, the network elements and their interconnections are usually organized into five hierarchical layers [3] or tiers, as shown

---

[3] VoIP GW translates time division multiplex (TDM) formatted voice signals into a real-time transport protocol (RTP) over a user datagram protocol (UDP) over IP packets.

[4] The GK controls one or more GWs and can interwork with the billing and management system of the PSTN network.

[5] The SG offers a mechanism for carrying SS7 signaling (mainly integrated services digital network [ISDN] user part [ISUP] and transaction capabilities application part [TCAP] messages over an IP network. IETF's RFC 2960 defines the stream control transmission protocol (SCTP) to facilitate this.

[6] Ethernet is the protocol of choice for local area networking (LAN). It has been standardized by the IEEE as its 802.3 protocol for media access control (MAC).

in Figure 1-1. The fifth layer contains end-office switches called CLASS-5 switches; examples are Lucent's 5ESSS, Nortel's DMS-100, and Siemens' EWSD. These switches provide connectivity to the end users via POTS or a black phone over the local copper plant or loop. In the United States, the regional Bell operating companies (RBOCs) such Verizon, Bell-South, SBC, and Qwest provide traditional POTS service to the residential and business customers (or users) in different local access and transport areas (LATAs).

Implementation of VoIP for CLASS-5 switch replacement for intra-LATA communication would require a breakdown of the PSTN switching system in a fashion similar to breaking down the mainframe computing model into a PC-based computing model. Therefore, one needs to think in terms of distributed implementation of control of call, service, and information transmission. Services that are hosted in the mainframe computer or in the CLASS-5 switches could be gradually migrated to server-based platforms and could be made available to end users inexpensively over IP-based networks.

VoIP can be implemented for inter-LATA (CLASS-4) and long-distance (both national and international, CLASS-3, -2, and -1) transmission of the voice signal as well. Figure 1-2 shows an implementation of long-distance voice transmission using the IP network for domestic long-distance services, assuming that the same company is allowed to offer both local and long-distance services in the LATAs that are being interconnected by an IP network. Here the network access from the terminal device (e.g., a black phone) can still be provided by a traditional CLASS-5 switch, but the inter-LATA transmission of a voice signal is offered over an IP network. The resulting architecture demands VoIP GWs to convert the TDM-formatted voice signal into IP packets at the ingress and vice versa at the egress. The VoIP GK controls call authentication, billing, and routing on the basis of the called phone number (E.164 address) and the IP address of the terminating VoIP GK. This is a classical implementation of VoIP service using the International Telecommunications Union's (ITU-T's) H.323 [4] umbrella protocols. The same architecture can be utilized or extended for international VoIP services, except that now the call-originating and call-terminating VoIP GWs would be located in two countries. Different countries usually deploy different voice signal companding schemes, use different formatting of voice signal compression mechanisms, and prefer different kinds of coding of signaling messages [5]. Therefore, the details of this type of design need to be carefully considered on a case-by-case basis.

## VoIP FOR ENTERPRISE CUSTOMERS

Some form of data communication network usually exists within any enterprise or corporation. These networks commonly utilize X.25, IP, frame relay (FR), and asynchronous transfer mode (ATM) technologies. However, recently, most of these networks have migrated to or are planning to use IP-based networks. Figure 1-3 shows such a network.

4

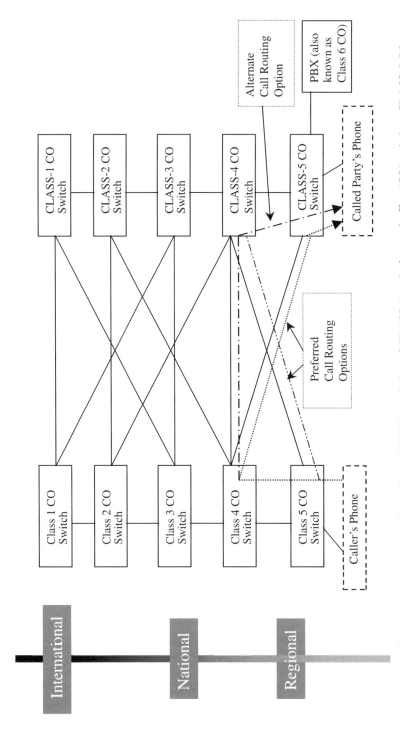

**Figure 1-1** The five-layer hierarchy of a traditional PSTN consisting of CLASS-1 to 5 of central office (CO) switches. CLASS-5 COs are commonly referred to as *end office* (EO) switches and CLASS-4 COs as *tandem* switches. The private automatic branch exchanges (PBX) are known as CLASS-6 COs as well. PBXs are used to provide traditional and enhanced PSTN/telephony services to business customers.

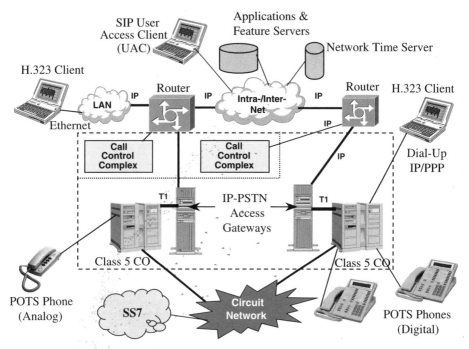

**Figure 1-2** A network configuration for supporting phone-to-phone, PC-to-phone, and PC-to-PC real-time voice telephony calls using a variety of VoIP protocols including the session initiation protocol (SIP) and H.323 Protocols. The call control complex hosts elements like the H.323 GK, SIP servers, Media Gateway Controller, SS7 SG, and so on, and contains all of the packet domain call control and routing intelligence. Applications and feature servers host the applications and services required by the clients. The network time server can be used for synchronizing the communicating clients with the IP-based Intranet/Internet.

For voice communications within the logical boundaries of an enterprise or corporation, VoIP can be implemented in buildings and on campuses both nationally and internationally. For small office home office (SOHO)-type services, multiple (e.g., two to four) derived phone lines with a moderately high (e.g., sub-T1 rate) speed would probably be sufficient. VoIP over the digital subscriber line (DSL; see, e.g., www.dsllife.com, 2001) channels or over coaxial cable can easily satisfy the technical and service requirements of the SOHOs. These open up new revenue opportunities for both telecom and cable TV service providers.

Most medium-sized and large enterprises have their own private branch exchanges (PBXs) for POTS/voice communication service, and hence they use sub-T1 or T1 rate physical connections to the telephone service providers' networks. They also have T1 rate and/or digital subscriber line (DSL)-type connections to facilitate data communications over the Internet. This current mode

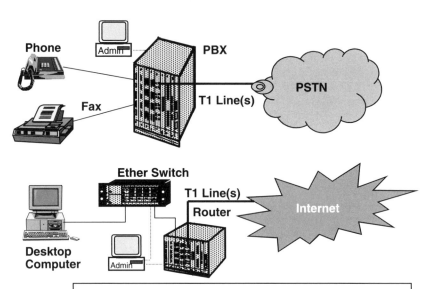

**Figure 1-3**   The elements and their interconnection in a traditional enterprise network.

of operation of separate data and voice communication infrastructures is shown in Figure 1-3. In an integrated communication environment, when VoIP is implemented, the same physical T1 and/or DSL link to the service provider's network can be used for both voice and data communications. The integrated infrastructure is shown in Figure 1-4. The details are discussed in the context of next-generation enterprise networks in Reference 6. One possible enterprise networking scenario that utilizes both IP and various types of DSL technologies for integrated voice, data, and video communications is shown in Figure 1-5 [6].

For very large corporations with nationwide branch offices and for multinational corporations with international offices, VoIP implementation may be preferable because such corporations may already have a large operational IP or overlay-IP network in place. The addition of VoIP service in such networks may need some incremental investments and has the potential to save the significant amount of money that is paid for leasing traditional telephone lines.

## FUNCTIONALLY LAYERED ARCHITECTURES

The traditional PSTN switching system is monolithic in nature, that is, almost all of its functionalities are contained in and integrated into one network element. This paradigm encourages vendors to use as many proprietary interfaces

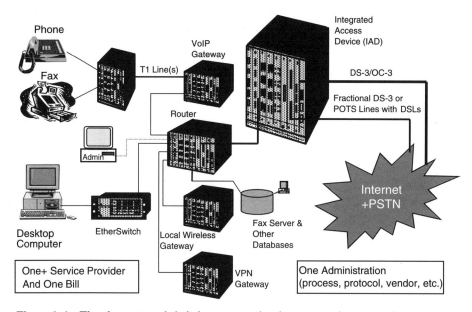

**Figure 1-4**  The elements and their interconnection in an emerging enterprise network.

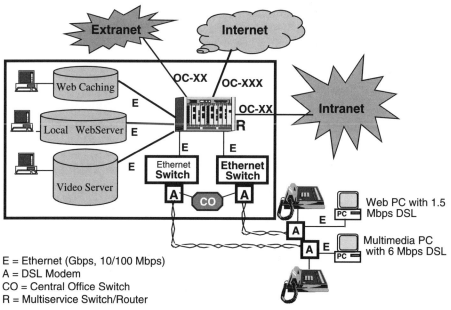

E = Ethernet (Gbps, 10/100 Mbps)
A = DSL Modem
CO = Central Office Switch
R = Multiservice Switch/Router

**Figure 1-5**  Next-generation enterprise networking using DSL- and IP-based technologies to support multimedia communications.

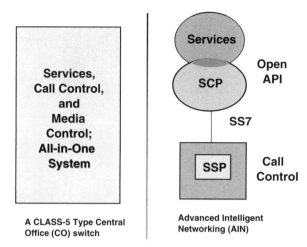

**Figure 1-6**   PSTN switch evolution using the AIN model. (Note: Elements such as SSP, SCP, SS7, and API are defined in the Glossary.)

and protocols as possible, as long as they deliver an integrated system that functions as per the specifications, which have been developed by Telcordia (formerly Bellcore, www.saic.com/about/companies/telcordia.html). However, this mode of operation also binds the PSTN service providers to the leniency of the vendors for (a) creation and management of services and (b) evolution and expansion of the network and system.

There have been many attempts in industry forums to standardize the logical partitioning of PSTN switching and control functions. Intelligent networking (IN) and advanced intelligent networking (AIN) were two such industry attempts. The AIN model is shown in Figure 1-6. AIN was intended to support at least the open application programming interface (API) for service creation and management so that the service providers could quickly customize and deliver the advanced call control features and related services that customers demand most often. However, many PSTN switch vendors either could not develop an open API or did not want to do so because they thought that they might lose market share. As a result, the objectives of the AIN efforts were never fully achieved, and PSTN service providers continued to be at the mercy of PSTN switch vendors for rolling out novel services and applications.

But then came the Internet revolution. The use of open/standardized interfaces, protocols, and technologies in every aspects of Internet-based computing and communications attempted to change the way people live and work. PSTN switching-based voice communication service was no exception. Many new standards groups were formed, and the standards industry pioneers such as ITU-T and IETF formed special study groups and work groups to develop standards for evolution of the PSTN systems. The purpose of all of these efforts was to make the PSTN system embrace openness not only in service cre-

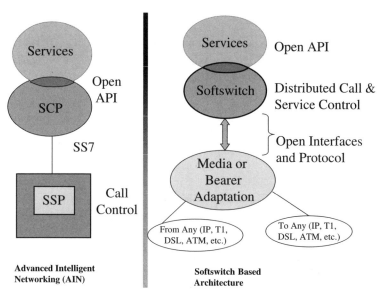

**Figure 1-7**  PSTN Switch evolution from using the AIN model to using softswitch-based architecture (Note: elements such as SSP, SCP, SS7, and API are defined in the Glossary.)

ation and management but also in switching and call control. As a result, the softswitch-based architecture was developed for PSTN evolution, as shown in Figure 1-7. A *softswitch* is a software-based network element that provides call control functions for real-time packet-voice (e.g., RTP over UDP over IP-based data streams) communications. This architecture enables incremental service creation and deployment, and encourages service innovation because it uses open APIs at the service layer. A softswitch uses a general-purpose computer server for hosting and executing its functions. Therefore, it supports some level of vendor independence that enables migration of PSTN switching system toward component-based architecture to support competitive procurement of network elements.

In general, a three-layer model, as shown in Figure 1-8, can be utilized for rolling out VoIP and other relevant enhanced IP-based communication services in an environment where the existing PSTN-based network elements have not yet fully depreciated. In this model, the elements on the right side represent the existing monolithic switching, transmission, and call control and feature delivery infrastructures. The elements on the left side represent a simplistic separation of bearer or media, signaling and control, and call feature delivery infrastructures. This separation paradigm closely follows the development of PC-based computing in contrast to mainframe-based computing. Therefore, it allows mixing and matching of elements from different vendors as long as the openness of the interfaces is maintained. In addition, it helps reduce system

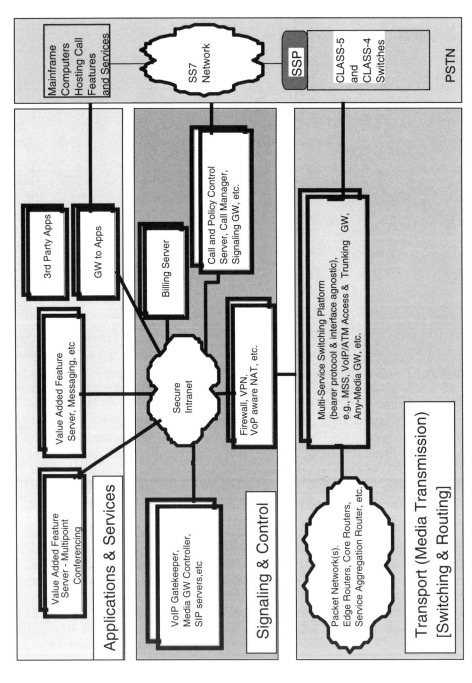

**Figure 1-8** A high-level three-layer generic functional architecture for PSTN evolution.

development and upgrading time cycles, facilitates rapid rollout of value-added services, and encourages openness in system-level management and maintenance mechanisms. For example, the call feature development and rollout time may be reduced from years in the traditional CLASS-5 switching system to a few weeks in the new paradigm. This opens up new revenue opportunities for the existing telecom and emerging competitive service providers. In addition, this architecture allows the telecom service providers to use both data transmission and voice transmission technologies in their networks to offer cost-effective transmission of data-grade voice and voice-grade data services [7], according to per customers' requirements.

The Multiservice Switching Forum (MSF at www.msforum.org, 2001) has recommended a more general multilayer model in their reference architecture implementation agreement (IA; available at www.msforum.org/techinfo/approved.shtml). This reference architecture is shown in Figure 1-9. This model essentially defines the functional elements or blocks in each layer and the reference interface points, with the objective of standardizing the functions and

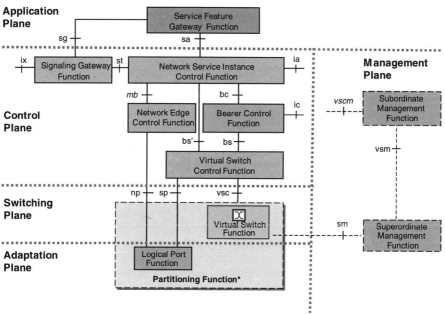

Notes:
- Italicized reference points are not considered open reference points for release 1.
- Bearer transport reference points are not shown.
|⊐| - Management functions overlaid on functional architecture
* The Partitioning Function maintains partition integrity between partitions of a partitioned entity.

**Figure 1-9**  MSF reference architecture for functional separation of call control, media adaptation, and application hosting. (Reproduced with permission from MSF's Release-1 Implementation Agreement, available at www.msforum.org/techinfo/approved.shtml, 2001.)

interface of the network elements. Networks developed using this architecture can support voice, data, and video services using existing and emerging transmission, switching, and signaling protocols.

A variety of other organizations are also working toward supporting similar open interface and protocol-based architectures for PSTN evolution. These include the International Softswitch Consortium (www.softswitch.org, 2001); various works groups (WGs) within IETF (www.ietf.org, 2001), including the PSTN/Internet interworking (PINT) and services in the PSTN/IN requesting Internet services (SPIRITS) WGs; various study groups (SGs) within the ITU-T (www.itu.int, 2001), including the SG11, which is currently working on evolution of the bearer independent call control protocols; and so on. Many websites also maintain up-to-date information on the latest development of IP telephony; for example, see the IP-Tel (www.iptel.org, 2001) website.

## ORGANIZATION OF THE BOOK

The rest of the book consists of eight chapters and three appendixes.

Chapter 2 discusses the existing and emerging voice coding and Internet technologies that are making the implementation of VoIP a reality. These include development of (a) low-bit-rate voice coding algorithms, (b) efficient encapsulation and transmission of packetized voice signal, (c) novel routing and management protocols for voice calls, (d) QoS delivery mechanisms for real-time traffic transmission over IP, and so on.

Chapter 3 presents the evolution of the VoIP call control and signaling protocols, beginning with the ones that assume that interworking with the traditional switch-based infrastructure (PSTN) is mandatory. I then discuss the procedures that use totally Internet-based protocols and the paradigm that uses server-router-based network architecture. Throughout the discussion, special emphasis is placed on the activities focused on making the PSTN domain-enhanced telephone call features available to IP domain clients like those using PC-based soft phones or hardware-based IP phones (e.g., session initiation protocol [SIP] phones) and vice versa.

Chapter 4 discusses a set of criteria that can be used to evaluate VoIP service irrespective of whether it is implemented in enterprise or residential networks. It appears that many of the PSTN domain reliability, availability, voice quality parameters, and call setup characteristics are either difficult to achieve or too costly to implement in an operational IP-based network unless one controls both the call-originating (or ingress) and call-terminating (or egress) sides of the network.

Chapter 5 reviews the architecture, hardware, and software elements of a recently developed testbed that can be used for subjective and objective evaluation of VoIP services. A special routing configuration of the access switch (e.g., a PBX) is utilized to route a telephone call over either a circuit-switched (PSTN) network or an internal IP-based network or Intranet. We used this testbed for evaluating the quality of transmission of real-time voice signal and dual-tone multifrequency (DTMF) digits over the Intranet with and without IP layer impairments. The NIST-Net impairment emulator (www.antd.nist.gov/itg/nistnet/, 2001) of the National Institute of Standards and Technology is utilized in the testbed to introduce impairments such as delay, delay jitter, packet loss, and bandwidth constraints.

Chapter 6 describes the advantages and techniques used in implementing the VoIP service in enterprise networks. It is possible to roll out easily the VoIP service in single-location enterprises. The network must be highly reliable and available to provide service even during interruption of the electric power supply and failure of one or more network servers. Customers should be able to use both IP phones and traditional POTS phones (with adapters) to make and receiver phone calls. Multimedia communication server and IP-PBX can be easily deployed in such a scenario, and these provide a real opportunity to integrate the corporate datacom and telecom infrastructures. For multisite medium-sized to large enterprises, implementation of access and transmission-level security and QoS may pose some challenges. However, many innovative solutions to these problems are available today.

Chapter 7 discusses a few technologies—such as DSL, cable TV, wireless local loop, and so on—and scenarios—for example, Web-based calling while surfing the Internet, flat-rate-based worldwide calling—in which the VoIP service can be implemented in public or residential networks. Introduction of the VoIP service in these networks would not only reduce operational and transmission costs, but also would accelerate deployment of many emerging networked host-based services. These next-generation services include unified communications, instant messaging and conferencing, Internet games, and others. I discuss the challenges of achieving PSTN-grade reliability, availability, security, and service quality using computer servers and IP-based network elements. Some reference implementation architectures and mechanisms are also mentioned in this chapter.

Chapter 8 illustrates how IP-based voice communication can be deployed in global enterprises. In traditional PSTN networks, various countries use their own version of the ITU-T standards for signaling or for bearer or information transmission. When IP-based networks, protocols, interfaces, and terminals (PCs, IP phones, Web clients, etc.) are used, unification of transmission, signaling, management, and interfaces can be easily accomplished. I discuss a possible hierarchical architecture for control of IP-based global communications for a hypothetical multinational organization.

In Chapter 9, based on experiences and experiments, I offer some recommendations to guide the implementation of VoIP services using any operational IP network. A list of the most challenging future research topics is then presented, followed by a discussion of industry efforts to resolve these issues.

Appendix A presents methodologies to measure the call progress time and to automate VoIP call setup for tests and measurements. Appendix B explains a technique that can be used to evaluate the bulk call handling performance of the VoIP GWs or IP-PSTN MGWs. Appendix C presents experimental evaluation of the quality of transmission of voice signal and DTMF digits under both impairment-free (i.e., typical PSTN) and impaired—that is, with added IP-level packet delay, delay jitter, and packet loss—networking conditions.

## EPILOGUE

Implementation of VoIP has reached a level of maturity that allows it to migrate easily from laboratory and prototype implementations to deployment in both residential and enterprise networks. Many new Internet protocol- and paradigm-based call control, signaling, and QoS guaranteeing techniques have been developed, and these are continuously evolving. Undoubtedly, they will accelerate the convergence of both voice and data services and the related network infrastructures, resulting in ubiquitous availability of the VoIP service.

## REFERENCES

1. G.177 Recommendation, Transmission Planning for Voiceband Services over Hybrid Internet/PSTN Connections, ITU-T, Geneva, 1999.
2. G. Thomsen and Y. Jani, "Internet Telephony: Going Like Crazy," IEEE Spectrum, Vol. 37, No. 5, pp. 52–58, May 2000.
3. R. J. Bates and D. W. Gregory, Voice and Data Communications Handbook, McGraw-Hill Book Companies, New York, 1998.
4. H.323 Recommendation, Packet-Based Multimedia Communications Systems, ITU-T, Geneva, 1999.
5. M. Tatipamula, and B. Khasnabish (Editors), Multimedia Communications Networks: Technologies and Services, Artech House Publishers, Boston, 1998.
6. B. Khasnabish, "Next-Generation Corporate Networks," IEEE IT Pro Magazine, Vol. 2, No. 1, pp. 56–60, January/February 2000.
7. B. Khasnabish, "Optical Networking Issues and Opportunities: Service Providers' Perspectives," Optical Networks Magazine, Vol. 3, No. 1, pp. 53–58, January/February 2002.

# 2

# TECHNOLOGIES SUPPORTING VoIP[1]

In this chapter, we discuss and review various standard and emerging coding, packetization, and transmission technologies that are needed to support voice transmission using the IP technologies. Limitations of the current technologies and some possible extensions or modifications to support high-quality—that is, near-PSTN grade—real-time voice communications services using IP are then presented.

## VOICE SIGNAL PROCESSING

For traditional telephony or voice communications services, the base-band signal between 0.3 and 3.4 KHz is considered the telephone-band voice or speech signal. This band exhibits a wide dynamic amplitude range of at least 40 dB. In order to achieve nearly perfect reproduction after switching and transmission, this voice-band signal needs to be sampled—as per the Nyquist sampling criteria—at more than or equal to twice the maximum frequency of the signal.

Usually, an 8 KHz (or 8000 samples per second) sampling rate is used. Each of these samples can now be quantized uniformly or nonuniformly using a predetermined number of quantization levels; for example, 8 bits are needed to support $2^8$ or 256 quantization levels. Accordingly, a bit stream of $(8000 \times 8)$ or 64,000 bits/sec (64 Kbps) is generated. This mechanism is known as the pulse code modulation (PCM) encoding of voice signal as defined in ITU-T's G.711 standard [1], and it is widely used in the traditional PSTN networks.

---

[1] The ideas and viewpoints presented here belong solely to Bhumip Khasnabish, Massachusetts, USA.

**Low-Bit-Rate Voice Signal Encoding**

With the advancement of processor, memory, and DSP technologies, researchers have developed a large number of low-bit-rate voice signal encoding algorithms or schemes. Many of these coding techniques have been standardized by the ITU-T. The most popular frame-based vocoders that utilize linear prediction with analysis-by-synthesis are the G.723 standard [2], generating a bit stream of 5.3 to 6.4 Kbps, and the G.729 standard [3], producing a bit stream of 8 Kbps. Both G.723 and G.729 have a few variants that support lower bit rate and/or robust coding of the voice signal. G.723 and G.723.1 coders process the voice signal in 30-msec frames. G.729 and G.729A utilize a speech frame duration of 10 msec. Consequently, the algorithmic portion of codec delay (including look-ahead) for G.723.1-based systems becomes approximately 37.5 msec compared to only 15 msec for G.729A implementations. This reduction in coding delay can be useful when developing a system where the end-to-end (ETE) delay must be minimized, for example, less than 150 msec to achieve a higher quality of voice.

An output frame of the G.723.1 coding consists of 159 bits when operating at the 5.3 Kbps rate and 192 bits in the 6.4 Kbps option, while G.729A generates 80 bits per frame. However, the G.729A coders produce three times as many coded output frames per second as G.723.1 implementations. Note that the amount of processing delay contributed by an encoder usually poses more of a challenge to the packet voice communication system designer.

Annex-B of G.729 or G.729B describes a voice or speech activity detection (VAD or SAD) method that can be used with either G.729 or its reduced complexity version, G.729A. The VAD algorithm enables silence suppression and comfort noise generation (CNG). It predicts the presence of speech using current and past statistics. G.729B allows insertion of 15-bit silence insertion descriptor (SID) frames during the silence intervals. Although the insertion of SID allows low-complexity processing of silence frames, it increases the effective bit rate. Consequently, although in a typical conversation, suppression of silence reduces the amount of data by almost 60%, G.729B generates a data stream of speed of little more than 4 Kbps.

The G.729A coder-decoder (CODEC) is simpler to implement than the one built according to the G.723.1 algorithm. Both designs utilize approximately 2K and 10K words of RAM and ROM storage, respectively, but G.729A requires only 10 MIPS, while G.723.1 requires 16 MIPS of processing capacity. The voice quality delivered by these CODECs is considered acceptable in a variety of network impairment scenarios. Therefore, most VoIP product manufacturers support G.723, G.729, and G.711 voice coding options in their products.

**Voice Signal Framing and Packetization**

PSTN uses the traditional circuit switching method to transmit the voice encoder's output (described above) from the caller's phone to the destination phone. The circuit switching method is very reliable, but it is neither flexible nor efficient for voice signal transmission, where almost 60% of the time the channel or circuit remains idle [4]. This happens either because of the user's silence or because the user—the caller or the party called—toggles between silence and talk modes.

In the packet switching method, the information (e.g., the voice signal) to be transmitted is first divided into small fixed or variably sized pieces called *payloads*, and then one or more of these pieces can be packed together for transmission. These packs are then encapsulated using one or more appropriate sets of headers to generate packets for transmission. These packets are called *IP packets* in the Internet, *frames* in frame relay networks, *ATM cells* in ATM networks [4], and so on. The header of each packet contains information on destination, routing, control, and management, and therefore each packet can find its own destination node and application/session port. This avoids the needs for preset circuits for transmission of information and hence gives the flexibility and efficiency of information transmission.

However, the additional bandwidth, processing, and memory space needed for packet headers, header processing, and packet buffering at the intermediate nodes call for incorporation of additional traffic and resource management schemes in network operations, especially for real-time communications services like VoIP. These are discussed in later chapters.

In G.711 coding, a waveform coder processes the speech signal, and hence generates a stream of numeric values. A prespecified number of these numeric values need to be grouped together to generate a speech frame suitable for transmission. By contrast, the G.723 and G.729 coding schemes use analysis-synthesis algorithms-based vocoders and hence generate a stream of speech fames, which can be easily adapted for transmission using packet-switched networks.

As mentioned earlier, it is possible to pack one or more speech frames into one packet. The smaller the number of voice or speech frames packed into one packet, the greater the protocol/encapsulation overhead and processing delay. The larger the number of voice or speech frames packed into one packet, the greater the packet processing/storing and transmission delay. Additional network delay not only causes the receiver's playout buffer to wait longer before reconstructing voice signal, it can also affect the liveliness/real-timeness of a speech signal during a telephone conversation. In addition, in real-time telephone conversation, loss of a larger number of contiguous speech frames may give the impression of connection dropout to the communicating parties. The designer and/or network operator must therefore be very cautious in designing the acceptable ranges of these parameters.

ITU-T recommends the specifications in G.764 and G.765 standards [5,6] for carrying packetized voice over ISDN-compatible networks. For voice transmission over the Internet, the IETF recommends encapsulation of voice frames using the RTP (RFC 1889) for UDP (RFC 768)-based transfer of information over an IP network. We discuss these in later sections.

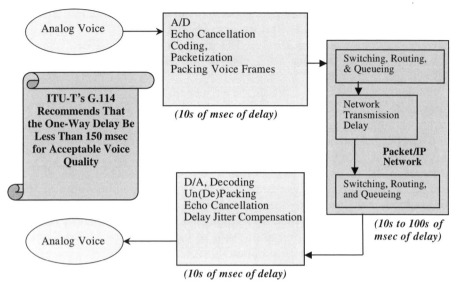

**Figure 2-1** A high-level packet voice transmission model.

## PACKET VOICE TRANSMISSION

A simple high-level packet voice transmission model is presented in this section. The schematic diagram is shown in Figure 2-1.

At the ingress side, the analog voice signal is first digitized and packetized (voice frame) using the techniques presented in the previous sections. One or more voice frames are then packed into one data packet for transmission. This involves mostly UDP encapsulation of RTP packets, as described in later sections. The UDP packets are then transmitted over a packet-switched (IP) network. This network adds (a) switching, routing, and queuing delay, (b) delay jitter, and (c) probably packet loss.

At the egress side, in addition to decoding, deframing, and depacking, a number of data/packet processing mechanisms need to be incorporated to mitigate the effects of network impairments such as delay, loss, delay jitter, and so on. The objective is to maintain the real-timeness, liveliness, or interactive behavior of the voice streams. This processing may cause additional delay. ITU-T's G.114 [7] states that the one-way ETE delay must be less than 150 msec, and the packet loss must remain low (e.g., less than 5%) in order to maintain the toll quality of the voice signal [8]. *end to end*

### Mechanisms and Protocols

As mentioned earlier, the commonly used voice coding options are ITU-T's G.7xx series recommendations (www.itu.int/itudoc/itu-t/rec/g/g700-799/),

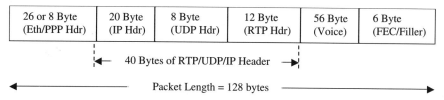

**Figure 2-2**    Encapsulation of a voice frame for transmission over the Internet.

three of which are G.711, G.723, and G.729. G.711 uses pulse code modulation (PCM) technique and generates a 64 Kbps voice stream. G.723 uses (CELP) technique to produce a 5.3 Kbps voice stream, and G.723.1 uses (MP-MLQ) technique to produce a 6.4 Kbps voice stream. Both G.729 and G.729A use (CS-ACELP) technique to produce an 8 Kbps voice stream.

Usually a 5 to 48 msec voice frame sample is encoded, and sometimes multiple voice frames are packed into one packet before encapsulating voice signal in an RTP packet. For example, a 30 msec G.723.1 sample produces 192 bits of payload, and addition of all of the required headers and forward error correction (FEC) codes may produce a packet size of ∼600 bits, resulting in a bit rate of approximately 20 Kbps. Thus, a 300% increase in the bandwidth requirements may not seem unusual unless appropriate header compression mechanisms are incorporated while preparing the voice signal for transmission over the Internet.

For example, a 7 msec sample of a G.711 (64 Kbps) encoded voice produces a 128 byte packet for VoIP application *including* an 18 byte MAC header and an 8 byte Ethernet (Eth) header (Hdr), as shown in Figure 2-2. Note that the 26 byte Ethernet header consists of 7 bytes of preamble, which is needed for synchronization, 12 bytes for source and destination addresses (6 bytes each), 1 byte to indicate the start of the frame, 2 bytes for the length indicator field, and 4 bytes for the frame check sequence.

The RTP/UDP/IP headers together add up to $20 + 8 + 12$, or 40 bytes of header. The IETF therefore recommends compressing the headers using a technique (as described in RFC 1144) similar to the TCP/IP header compression mechanism. This mechanism, commonly referred to as *compressed RTP* (CRTP, RFC 2508), can help reduce the header size from (12 to 40) bytes of RTP/UDP/IP header to 2 to 4 bytes of header. This can substantially reduce the overall packet size and help improve the quality of transmission.

Note that the larger the packet, the greater the processing, queueing, switching, transmission, and routing delays. Thus, the total ETE delay could become as high as 300 msec [8], although ITU-T's G.114 standard [7] states that for toll-quality voice, the one-way ETE delay should be less that 150 msec. The mean opinion score (MOS) measure of voice quality is *usually* more sensitive to packet loss and delay jitter than to packet transmission delay. Some information on various voice coding schemes and quality degradation because

| Ver. (4) | IHL (4) | Type of Service (8 Bits) | Total Length (16 Bits) |
|---|---|---|---|
| Fragment Offset (12 Bits) | | Flags (4 Bits) | ID (16 Bits) |
| Time to Live (8 Bits) | | Protocol (8 Bits) | Header Checksum (16 Bits) |
| Source Address (32 Bits) | | | |
| Destination Address (32 Bits) | | | |
| Options (24 Bits) | | | Padding (8) |

**IHL:** Internet Header Length

**Figure 2-3**   IP version 4 (IPv4) header format. (Source: IETF's RFC 791.)

of transmission can be found at the following website: www.voiceage.com/products/spbybit.htm

The specification of the IETF's (at www.ietf.org) Internet protocol version 4 (IPv4) is described in RFC 791, and the format of the header is shown in Figure 2-3. IP supports both reliable and unreliable transmission of packets. The transmission control protocol (TCP, RFC 793; the header format is shown in Figure 2-4) uses window-based transmission (flow control) and explicit acknowledgment mechanisms to achieve reliable transfer of information. UDP (RFC 768; the header format is shown in Figure 2-5) uses the traditional "send-and-forget" or "send and pray" mechanism for transmission of packets.

There is no explicit feedback mechanism to guarantee delivery of information, let alone the timeliness of delivery. TCP can be used for signaling, parameter negotiations, path setup, and control for real-time communications like VoIP. For example, ITU-T's H.225 and H.245 (described below) and IETF's domain name system (DNS) use the TCP-based communication pro-

| Source Ports (16 Bits) | | Destination Ports (16 Bits) |
|---|---|---|
| Sequence Number (32 Bits) | | |
| Ack. Number (32 Bits) | | |
| Off set (4) | Rsr vd (6) | U-A-P-R-S-F (1 Bit each) | Window (16 Bits) |
| Checksum (16 Bits) | | Urgent Pointer (16 Bits) |
| Options (16 Bits) | | Padding (16 Bits) |

**Control Bits** ⇒ **U:** Urgent Pointer; **A:** Ack.; **P:** Push function; **R:** Reset the connection; **S:** Synchronize the sequence number; **F:** Finish, means no more data from sender

**Figure 2-4**   TCP header format. (Source: IETF's RFC 793.)

UDP Header
= 8 Bytes or 64 Bits

RTP Header
= 12 Bytes or 96 Bits

| Src-Port (16 Bits) | Dest-Port (16 Bits) |
|---|---|
| Length (16 Bits) | Checksum (16 Bits) |
| Data (32 Bits) | |

| Xten.(1) | CC (4) | M (1) | PT (7) |
|---|---|---|---|
| Ver. (2) | Pad (1) | | Seq. No. (16 Bits) |
| Timestamp (32 Bits) | | | |
| Sync. Source ID (32 Bits) | | | |

PT: Payload Type; M: Marker
CC: Contributing Source Count
Xten.: Extension

**Figure 2-5**   UDP and RTP header formats. (Source: IETF's RFC 768 and 1889.)

tocol. UDP can be used for transmission of payload (traffic) from sources generating real-time packet traffic. For example, ITU-T's H.225, IETF's DNS, IETF's RTP (RFC 1889; the header format is shown in Figure 2-5), and the real-time transport control protocol (RTCP, RFC 1890) use UDP-based communications.

ITU-T's H.323 uses RTP for transfer of media or bearer traffic from the calling party to the destination party, and vice versa once a connection is established. RTP is an application layer protocol for ETE communications, and it does not guarantee any quality of service for transmission. RTCP can be used along with RTP to identify the users in a session. RTCP also allows receiver report, sender report, and source descriptors to be sent in the same packet. The receiver report contains information on the reception quality that the senders can use to adapt the transmission rates or encoding schemes dynamically during a session. These may help reduce the probability of session-level traffic congestion in the network.

Even though IPv4 is the most widely used version of IP in the world, the IETF is already developing the next generation of IP (IPv6, RFC 1883; the header format is shown in Figure 2-6). It is expected [9] that the use of IPv6 will alleviate the problems of security, authentication, and address space limitation (a 128 bit address is used) of IPv4. Note that proliferation of the use of the dynamic host control protocol (DHCP, RFC 3011) may delay widespread implementation of the IPv6 protocol.

Although there are many protocols and standards for control and transmission of VoIP, ITU-T's H.22x and H.32x recommendations (details are available at www.itu.int/itudoc/itu-t/rec/h/) are by far the most widely used. The H.225 standard [10] defines Q.931 protocol-based call setup and RAS (registration, administration, and status) messaging from an end device/unit or terminal device to a GK. H.245 [11] defines in-band call parameter (e.g., audiovisual mode and channel, bit rate, data integrity, delay) exchange and

| Ver. (4) | TOS or DS (8 Bits) | Flow Label (20 Bits) | | |
|---|---|---|---|---|
| Payload Length (16 Bits) | | NH (8 Bits) | Hop Count (8 Bits) | |
| Source Address (128 Bits) | | | | |
| Destination Address (128 Bits) | | | | |

**NH**: Next Header; can be used for supporting **Authentication**,
   **Security**, **Multicasting** and **Mobility**
**TOS** or **DS**: Type of Service or Diff-Serv byte; can be used to
   support the Quality of Service (**QoS**) requirements

**Figure 2-6**  IP version 6 (IPv6) header format. (Source: IETF's RFC 1883.)

negotiation mechanisms. H.320 defines the narrowband video telephony system and terminal; H.321 defines the video telephony (over an asynchronous transfer mode [ATM]) terminal; H.322 defines the terminal for video telephony over a LAN where the QoS can be guaranteed; H.323 [12] defines a packet-based multimedia communications system using a GW, a GK, a multipoint control unit (MCU), and a terminal over a network where the QoS cannot be guaranteed; and H.324 defines low-bit-rate multimedia communications using a PSTN terminal. Over the past few years, a number of updated versions of H.323 have appeared. H.235 [13] defines some relevant security and encryption mechanisms that can be applied to guarantee a certain level of privacy and authentication of the H-series multimedia terminals. H.323v2 allows fast call setup; it has been ratified and is available from many vendors. H.323v3 provides only minor improvements over H.323v2. Currently, work is in progress on H.323v4 and H.323v5. Because of its widespread deployment, H.323 is currently considered the legacy VoIP protocol. Figure 2-7 shows the protocol layers for real-time services like VoIP using the H.323 protocol.

Other emerging VoIP protocols are IETF's session initiation protocol (SIP, RFC 2543), media gateway control protocol (MGCP, RFC 2805), and IETF's Megaco (RFC 3015)/ITU-T's H.248 standards. SIP defines call-processing language (CPL), common gateway interface (CGI), and server-based applets. It allows encapsulation of traditional PSTN signaling messages as a MIME attachment to a SIP (e-mail) message and is capable of handling PSTN-to-PSTN calls through an IP network. MGCP attempts to decompose the call control and media control, and focuses on centralized control of distributed gateways. Megaco is a superset of MGCP in the sense that it adds support for media control between TDM (PSTN) and ATM networks, and can operate over either UDP or TCP. Figure 2-8 shows the protocol layers for VoIP call control and signaling using the SIP protocol. Figure 2-9 depicts the elements of MGCP and Megaco/H/248 for signaling and control of the media gateway. The details of these protocols are discussed in the next chapter.

| Applic-ation | Call Establishment and Session Control | | | | | |
|---|---|---|---|---|---|---|
| Present-ation | ADDRES-SING | | PRESENTATION (Media Transmission) | | | |
| | | | DTMF | Codec (G.7XX) | ADDRESSING | |
| Session | DNS | H.225 (Q.931) [helps exchange of control info. for H.323 services ] | H.245 [helps in-band message exchange for call control ] | RTP (rfc 1889) - puts sequence number, packet type, timestamp; RTCP (rfc1890) - monitors QoS & passes this info. back & forth | H.225 - RAS [terminals use it to access the GK] | DNS - rfc 2065, 2136/37 [terminals use it to discover the IP address of GK] |
| Transport | TCP (reliable), RFC 793 | | | UDP (unreliable), RFC 768 | | |

**Figure 2-7**  Protocol layers for H.323v1-based real-time voice services using the IP. RAS: registration, administration, status; GK: gatekeeper. Note that H.323v2 allows fast call setup by using H.245 within Q.931, and can run on both UDP and TCP.

For survivability, all of these protocols must interwork gracefully with H.323- and/or SIP-based VoIP systems. Industry forums like the International Multimedia Telecommunications Consortium (IMTC, at www.imtc.org, 2001), the Multiservice Switching Forum (MSF, at www.msforum.org, 2001), the Open Voice over Broadband Forum (OpenVoB, at www.openvob.com, 2001), and the International Softswitch Consortium (www.softswitch.org, 2001) are actively looking into these issues, and proposing and demonstrating feasible solutions. OpenVoB is initially focusing on packet voice transmission over digital subscriber lines (DSL). Depending on the capabilities of the DSL modem

| Applic-ation | Call Control and Signaling | | | | | |
|---|---|---|---|---|---|---|
| Present-ation | ADDRES-SING | | PRESENTATION (Media Transmission) | | | |
| | | | DTMF | Codec (G.7XX) | ADDRESSING | |
| Session | DNS | SIP Signaling Messages [Register, Ack, Options, Cancel, Bye, etc.] | | RTP (rfc 1889) - puts sequence number, packet type, timestamp; RTCP (rfc1890) - monitors QoS & passes this info. back & forth | TRIP (rfc2871) [terminals can use it to locate a server] | DNS - rfc 2065, 2136/37 [terminals use it to discover IP address] |
| Transport | TCP (reliable), RFC 793 | | | UDP (unreliable), RFC 768 | | |

**Figure 2-8**  Protocol layers for SIP-based real-time voice services using the IP.

| Applic-ation | Call Signaling/Routing and Gateway Control | | | | | |
|---|---|---|---|---|---|---|
| Present-ation | Addressing | API | | | PRESENTATION (Media Transmission) | | ADDRES SING |
| | | | | | DTMF | Codec (G.7XX) | |
| Session | DNS | Trans-action Cont-rol | Timer Manage ment | Command Control | RTP (rfc 1889) - puts sequence number, packet type, timestamp; RTCP (rfc1890) - monitors QoS & passes this info. back & forth | DNS - rfc 2065, 2136/37 [terminals use it to discover the IP address of a device] |
| Transport | TCP (RFC 793) | UDP (in MGCP); TCP and/or UDP (in H.248/Megaco) | | | UDP (RFC 768) | |

**Figure 2-9**   Protocol layers for MGCP and Megaco/H.248-based real-time voice services using IP.

or the integrated access device (IAD), it is possible to use either voice over ATM or VoIP over ATM to support the VoDSL service. If VoIP is used for VoDSL, then it is highly likely that the IAD has to support SIP or MGCP (migrating to H.248/Megaco)-based clients as voice terminals.

Finally, Figure 2-10, shows various existing and emerging services that use IP as the network layer protocol along with their RFC numbers. A detailed

| OSI | RFCs and Other Implementations | | | | | | | |
|---|---|---|---|---|---|---|---|---|
| Applic-ation | | | E-Mail | Terminal | | Client/Srvr | Net. Mngmnt | |
| Present-ation | HTTP | FTP | SMTP | Telnet | DNS | NFS RFC 1014/ 57/94 | SNMP RFC 1157 (v1) RFC 1901/10(v2) RFC 2271/75(v3) | |
| Session | RFC 2068 | RFC 959 | RFC 821 | RFC 854 | RFC 1034/35 | | | |
| Transport | TCP (reliable) RFC 793 | | | | UDP (unreliable) RFC 768 | | | |
| Network | Adrs. Resolution Protocols R/ARP RFC 826/903 | | IP (v4 is RFC 791, v6 is RFC 2460); over e.g., ATM, FR, TDM, X.25 "Links" | | | | ICMP RFC 792 | |
| Link | IEEE 802.2 (LLC), IEEE 802.1p (eight priorities) and IEEE 802.1Q (4-Byte tag, VLAN) Ethernet (IEEE 802.3), TokenRing, LAN, MAN, RFC 894, 1202, 1042, etc. | | | | | | | |
| Physical | Physical Media Twisted Pair, Coax, Fiber, Wireless Channels, etc. | | | | | | | |

**Figure 2-10**   The internet protocol layers.

description of each of these RFCs can be found at IETF's website (www.ietf. org/rfc.html).

### Packet Voice Buffering for Delay Jitter Compensation

In packet-switched networks, buffering serves many useful and repugnant purposes. At the access domain, a buffer provides temporary storage space for packets before they are routed to the appropriate transport network. The amount of delay suffered by packets at the access networks depends on buffer size, traffic density, and packet priority if it is supported.

At the transport domain, buffering is needed to support proper routing and multiplexing of packets, which help improve utilization of network links. Packet transport delay depends on the following factors:

a. Buffer sizes at ingress and egress of transport link(s). These depend on link transmission capacity; for example, T1, T3, OC-3, and OC-12 links have 1.544, 45.736, 155.52, and 622.08 Mbps of capacity, respectively [14];

b. Packet propagation time. This depends on the physical length of the transmission link—for example, 5.0 µsec/km if the signal (electrical or optical) travels at a speed of $2 \times 10^8$ m/s;

c. Transmission capacity or bandwidth of the link, as mentioned earlier; and

d. Packet storing and header processing delays at the intermediate nodes.

Finally, at the packet delivery domain, the packets that arrive earlier than the expected time need to be stored temporarily before being delivered for playout (for voice) or display (for video). Similarly, packets that arrive later than the expected time may need to be stored for a certain amount of time. The *expected time* is the mean or average value of a large sample of observed values of packet transfer delay from the source port or node to the destination port or node under *nominal* network traffic load.

For non-real-time data communications, delayed packets can be stored for an indefinite amount of time at local buffers. For real-time applications like VoIP service, delayed packets may become useless after a prespecified amount of time. The *delay jitter buffer* holds these "precocious" and delayed packets in an attempt to neutralize the effects of packet interarrival jitter. This helps maintain the real-timeness or liveliness of real-time communication over packet-switched networks. The delay jitter buffer must be neither too small nor too large. If it is too small, it will not serve its purpose, and if it is too large, it may remain filled with useless packets (i.e., that cannot be sent to the playout buffer) for a log time.

Ideally, the size of the delay jitter buffer should vary from a few to several speech frames, and its threshold—to prevent underflow and overflow—should adapt to changing network traffic conditions. Consequently, the additional delay due to this buffer would not adversely affect voice quality.

As defined in IETF's RFC 1889, the interarrival jitter $J$ is the mean deviation of the difference $D$ in packet spacing at the destination compared to the source for a pair of packets. This is equivalent to the difference in the relative transit time for the two packets. The *relative transit time* is the difference between a packet's RTP timestamp and the receiver's clock at the time of arrival, measured in the same units. For example, if $Si$ is the RTP timestamp from packet $i$ and $Ri$ is the time of arrival in RTP timestamp units for packet $i$, then for two packets $i$ and $j$, $D$ is expressed as $D(i, j) = [(Rj - Ri) - (Sj - Si)] = [(Rj - Sj) - (Ri - Si)]$.

The interarrival jitter is calculated regularly as each data packet $i$ is received from the source SSRC $- n$, using this difference $D$ for that packet and the previous packet $i - 1$ in order of arrival (not necessarily in sequence) using the following first-order estimator:

$$J_i = J_{i-1} + \frac{|D_i - D_{i-1}| - J_{i-1}}{16}$$

The gain parameter $1/16$ is used because it provides a reasonable noise reduction ratio while maintaining a reasonable rate of convergence. The current value of $J$ is sampled when a reception report is issued.

### QoS Enforcement and Impairment Mitigation Techniques

The parameters that define the service level agreement (SLA) and QoS are tightly coupled with the applications they are supporting. For example, the QoS parameters for access and transport networks may be different; the QoS parameters for real-time and non-real-time communications are different, and so on. In order to maintain liveliness or a certain degree of interactivity, real-time traffic (data or packet) must reach the destination *within* a preset time interval (*delay*) *with* some tolerance (*jitter*). Otherwise, it would be considered lost traffic. Some critical (important) non-real-time traffic, such as topology and routing related information, is loss-sensitive. Entire network could collapse if these packets are lost!

The following two techniques can be used to satisfy the QoS requirements of real-time packet traffic: (a) overallocation of bandwidth and (b) prioritization and scheduling of packets for service. Although these methods can be deployed independently, the best results can probably be achieved when a combination of them is used. The most effective solution would include the use of *preventive* and/or proactive traffic management schemes at *Access*, Network, and Nodal operation levels, and the use of *reactive* traffic management schemes at *Nodal*, Access, and Network operation levels [15].

*Preventive Mechanisms*  Preventive control mechanisms at the Access level include the use traffic descriptor, traffic contract, conformance testing, and so on to exercise control.

At the Network level, overprovisioning of link capacity, sharing and/or spreading of traffic across various routes to a destination (most useful for non-real-time traffic) can be used.

At the Nodal (queueing) operations, judicious use traffic shaping at the intermediate nodes can be used to exercise control. The ATM technology has well-defined mechanisms built into it to support most of these features [4,16–18]. The IETF is making efforts to incorporate similar traffic management mechanisms in IP. Examples are the activities related to MPLS, integrated services (IntServ), differentiated services (DiffServ), and so on in IETF's transport area working groups (www.ietf.org/html.charters/wg-dir.html# Transport_Area).

*Reactive Mechanisms*  Reactive control mechanisms, at the Nodal (queueing) operations include discarding packets if the queue size is *growing quickly* and the incoming packets are neither important nor urgent. At the Access level, the packets can be marked, for example by using the IP type of service (TOS) byte, or discarded on the basis of port or connection type if oversubscription *persists* in a session.

In the Network, the traffic flow rate can be controlled in physical and virtual connections using the route congestion information flowing back and forth. The criterion here is that the response or reaction time must be *fast* enough for the *control* to be effective and useful. Table 2-1 shows one possible method of categorizing control, signaling, and media traffic for supporting VoIP service. The corresponding mechanism for multipriority queueing and servicing of packets is presented in Figure 2-11.

The following mathematical formulation (see, e.g., Ref. 19) can be used for dimensioning the size of each of the buffers or queues shown in Figure 2-11.

**TABLE 2-1    An Example of Traffic Prioritization for Supporting Real-Time VoIP**

| Type of Information | Emission Priority | Discard Priority | Comments |
|---|---|---|---|
| ~ Urgent and important | Low | *Mostly* nondiscardable; (occasionally set loss priority [LP] = 0) | Session-level control and signaling traffic |
| Urgent and important | Medium | Nondiscardable; (LP = 0) | Network management and control traffic |
| Urgent and ~ important | High | Discardable (LP = 1) | Bearer or media traffic (e.g., voice or speech signal) |

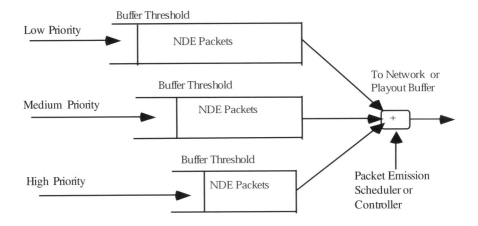

NDE: Non -Discard Eligible

**Figure 2-11**   An example of multipriority queueing for supporting real-time VoIP.

$$Queue\ Size\ (MTUs) = \frac{[\ln(P_{\text{loss}}) - \ln(\rho(1 - \rho))]F}{[(MTU)\gamma]}$$

$$\gamma = \frac{[2(\rho - 1)]}{[\rho C_a^2 + C_s^2]}$$

where

$P_{\text{loss}}$ is the probability of loss of MTU (message transmission unit); for example, $10^{-4}$

$\rho$ is the traffic intensity or link utilization (e.g., 0.85)

$C_a^2$ is [variance/(mean or average)$^2$] of the packet ($F$ MTUs) arrival process

$C_s^2$ is [variance/(mean or average)$^2$] of the packet service process

Typical values or ranges of the buffer dimensioning parameters are as follows:

- The probability of packet loss $P_{\text{loss}}$ is $10^{-6}$ for real-time (high- or medium-priority traffic) and $10^{-4}$ for non-real-time (medium- or low-priority traffic) traffic.
- The utilization $\rho$ varies from 0.80 to 0.99.
- The MTU is usually 64 bytes for high-priority traffic, 128 bytes for medium-priority traffic, and 512 bytes or higher for low-priority traffic.
- The coefficient of variation $C_a^2$ for the arrival process varies (from experiments) from 1.0 to 1.5 for high-priority traffic, from 1.0 to 3.24 for medium-priority traffic, and from 2.4 to 5.00 for low-priority traffic.

- The coefficient of variation $C_s^2$ for the service process varies (from experiments) from 0.8 to ~1.0 for high-priority traffic, from 0.5 to 0.6 for medium-priority traffic, and from 0.2 to 0.3 for low-priority traffic.

An example of buffer size computation for medium-priority traffic is as follows: Let $\{P_{loss}, \rho, MTU, C_a^2, C_s^2\} = \{10^{-6}, 0.95, 128, 3.24, 0.60\}$, and using the previously cited formulations, the buffer size become approximately 50 KB, as shown below.

$$Q_{size} \text{ (bytes)} = \frac{[\ln_e(10^{-6}) - \ln_e(0.95(1 - 0.95))] \times 128}{\gamma}$$

$$\gamma = \frac{2 \times (0.95 - 1.0)}{(0.95 \times 3.24) + 0.6} = \frac{-0.10}{3.678} = -0.02719$$

$$Q_{size} \text{ (bytes)} = \frac{[\ln_e(10^{-6}) - \ln_e(0.0475)] \times 128}{-0.02719}$$

$$Q_{size} \text{ (bytes)} = \frac{[-13.81551 + 3.047] \times 128}{-0.02719}$$

$$Q_{size} = \frac{-10.76851 \times 128}{-0.02719} = 50 \text{ KB}$$

This 50 KB of buffer space is equivalent to 270 msec of *maximum* delay on a T1 (1.544 Mbps) link. To minimize the maximum queueing delay, the *network design* should consider *minimizing* the number of *active* nodes crossed from source to destination. Consequently, the concept of *virtual (private) networking* (VPN) comes into the picture.

*Future Directions*   A number of efforts are currently underway in the standards organizations that incorporate mechanisms to support QoS in IP-based networks. In general, for IP, packet prioritization using the TOS byte, queue dimensioning, and scheduling (as discussed earlier) including the weighted fair queueing (WFQ) technique—as proposed by some vendors—can be used. IETF's DiffServ uses the TOS byte in IPv4 or the DS byte in IPv6 to define the per hop behavior (PHB) of traffic, the traffic marker in PHB, and so on. The ETE QoS can also be defined by IETF's IntServ, and using the resource reservation protocol (RSVP, RFC 2205), it can be signaled from source to destination. This mechanism is useful especially in network backbones. IETF is currently addressing the scalability of this mechanism.

Another evolving mechanism for supporting ETE QoS is the multiprotocol label switching (MPLS) technique. In MPLS a 32-bit label, for example, is added in the IP packet to maintain the desired ETE QoS. Both IETF (see, e.g., RFCs 3031, 3032, 3035, 3036, etc. at www.ietf.org/rfc.html, 2001) and the MPLS forum (see www.mplsforum.org, 2001, for details) are currently considering various techniques for distribution of labels (LDP) and for setting up a label switched path (LSP) in an IP network. In order to accelerate the deployment of VoIP in multiservice networks, the MPLS forum has recently released its implementation agreement to support real-time voice transmission over MPLS (VoMPLS). Interested readers can download that document from the MPLS forum's website (www.mplsforum.org, 2001).

## EPILOGUE

Recent advances in computing, DSP, and memory technologies have allowed the development of many sophisticated low-bit-rate speech coding algorithms. These processor- and memory-intensive techniques further delay the digitization and packetization of the voice signal, which commonly results in degradation of voice quality. Additional hardware-based echo-canceller and higher-speed transmission mechanisms are generally required to improve voice quality in such scenarios.

As shown in Figure 2-2, IP-based transmission of a digitized voice signal for (real-time) telephony service requires the addition of multiple levels of encapsulation overheads. This causes an increase in the bandwidth requirement for a voice session (or call) unless a header compression mechanism is utilized. In addition, when both real-time voice and non-real-time data traffic are transported over the same IP network, proactive traffic management techniques need to be incorporated in the routers and switches in order to maintain timely—evenly spaced and with low loss—transmission of voice traffic. Otherwise, the voice quality will be degraded. Many of the existing Internet protocols and networking techniques are currently being modernized (see, e.g., www.internet2.edu, www.ipv6forum.org, www.mplsforum.org, etc.) to support directly the transmission of real-time traffic for audio and video applications.

## REFERENCES

1. G.711 Recommendation, Pulse Code Modulation (PCM) of Voice Frequencies, ITU-T, Geneva, 1988.
2. G.723.1 Recommendation, Dual Rate Speech Coder for Multimedia Telecommunication Transmitting at 5.3 and 6.4 kbit/s, ITU-T, Geneva, 1996.
3. G.729A Recommendation. (1996). Coding of Speech at 8 kbit/s Using Conjugate Structure Algebraic Code Excited Linear Prediction (CS-ACELP)—Annex A: Reduced Complexity 8 kbit/s CS-ACELP Speech Codec, ITU-T, Geneva, 1996.
4. D. Minoli and E. Minoli, Delivering Voice over Frame Relay and ATM, John Wiley & Sons, New York, 1998.
5. G.764 Recommendation, Voice Packetization—Packetized Voice Protocols, ITU-T, Geneva, 1990.

6. G.765 Recommendation, Packet Circuit Multiplication Equipment, ITU-T, Geneva, 1992.

7. G.114 Recommendation, One-Way Transmission Time, ITU-T, Geneva, 1996.

8. IEEE, IEEE Network Magazine, IEEE Press/Publishers, New York, Vol. 12, No. 1, January/February 1998.

9. C. Huitema, IPv6—The New Internet Protocol, Prentice-Hall, Upper Saddle River, New Jersey, 1998.

10. H.225 Recommendation, Call Signaling Protocols and Media Stream Packetization for Packet-Based Multimedia Communication Systems, ITU-T, Geneva, 2000.

11. H.245 Recommendation, Control Protocol for Multimedia Communication, ITU-T, Geneva, 1998.

12. H.323 Recommendation, Packet-Based Multimedia Communications Systems, ITU-T, Geneva, 1999.

13. H.235 Recommendation, Security and Encryption for H-Series (H.323 and Other H.245-Based) Multimedia Terminals, ITU-T, Geneva, 1998.

14. W. Stallings, Data and Computer Communications, 6th edition, Prentice Hall, Upper Saddle River, New Jersey, 2000.

15. B. Khasnabish and R. Saracco (editors), "Intranets: Technologies, Services, and Management," IEEE Communications Magazine, Vol. 35, No. 10, pp. 78–121, October 1997.

16. ATM Forum, ATM Traffic Management Specifications, Version 4.0, 1996.

17. ATM Forum, Voice and Telephony Over ATM to the Desktop Specification, AF-VTOA-0083.001, 1999.

18. M. Tatipamula and B. Khasnabish (editors), Multimedia Communications Networks: Technologies and Services, Artech House Publishers, Boston, 1998.

19. M. C. Springer and P. K. Maken, "Queueing Models for Performance Analysis: Selection of Single Station Models," European Journal of Operational Research, #58, 1991.

# 3

# EVOLUTION OF VoIP SIGNALING PROTOCOLS[1]

This chapter reviews the existing and emerging VoIP signaling and call control protocols. In PSTN networks, ISUP (ISDN user part) and TCAP (transaction capabilities application part) messages of the SS7 protocol [1] are commonly used for call control and interworking of services.

The first generation (released in 1996) of VoIP signaling and media control protocols, such as ITU-T's H.225/H.245—defined under ITU-T's H.323 umbrella protocol [2]—was intended to offer LAN-based real-time VoIP services. These protocols already had the proper ingredients (such as support of ISUP messaging for call control) to support interworking with PSTN networks as well. Consequently, there was a flurry of networking activities to deliver VoIP services in LAN or within enterprises and to offer long-haul (inter-LATA and international) transport of VoIP. The latter is also known as *cheap and wireless quality long-distance voice service over wireline network using IP*. However, the telecom service providers found the following two problems with version 1 of the H.323 protocol:

    a. Many of the desired and advanced PSTN-domain call features and services could not be easily implemented using H.323v1 because of its lack of openness (i.e., all of the procedures are internally defined), and

    b. Scalable implementation was neither feasible nor cost-effective because it needed call state full proxies.

---

[1] The ideas and viewpoints presented here belong solely to Bhumip Khasnabish, Massachusetts, USA.

These problems motivated ITU-T to release the second version of H.323 in 1998. H.323-v2 supports lightweight call setup—runs over UDP instead of using multiple TCP sessions per call—and declares many of the mandatory features and protocols of H.323v1 to be optional [3]. But in 1999, IETF released the first version of its Internet paradigm, Web protocol (i.e., HTTP), and well-defined semantics-based session initial protocol (SIP, RFC 3261) for VoIP call control, and service (a superset of the PSTN domain) creation and management. In addition, SIP supports call stateless proxies and allows traversal of call states over many proxy hops [4,5]. These make scalable implementation of VoIP more feasible than was possible using H.323.

The race to catch up continued. ITU-T announced versions 3 and 4 of H.323 and then certified H.323v4 in late 2000. H.323v4 supports the following features: (a) extensive support of UDP, SCTP (defined later in this chapter), and making H.245 optional; (b) enhanced support of security as defined in H.235v2; (c) support of H.323 (URL) for a network-based presence and instant messaging; and (d) support of tunnel-based signaling like ISUP, Q.SIG, and so on and HTTP commands and stimulus-based call control.

IETF is also working on extending the service creation, security, and call routing features of SIP (RFCs 3261, 3262, 3263, 3264, 3265, and 3266). Some of these features are (a) instant messaging and presence management, (b) advanced call routing and messaging features, and (c) support of SIP/SDP/RTP message traversal over network address translation (NAT) and firewall devices.

In parallel to the above-mentioned activities related to H.323v2 (and beyond) and SIPv2, researchers at Cisco, Level 3 Communications, and Telcordia developed a call/media control architecture for the next-generation (packet-based) network that supports both IP telephony and evolution of PSTN from a monolithic system to one that supports distributed call processing. That architecture enables physical separation of call control intelligence that resides in the media gateway controller (MGC) from the media-adaptation/translation gateways (MGs). It also recommends a protocol called MGCP (media gateway control protocol, RFC 2705, 1999), which was the result of a merger of SGCP (simple gateway control protocol) and IPDC (IP device control) protocol. MGCP supports PSTN evolution by allowing interworking with circuit-switched networks and devices (analog and digital POTS phones) via the following predefined endpoints: (a) access and residential GWs, and integrated network access server and VoIP GWs; (b) GWs supporting ISUP and multifrequency-type trunks; and (c) announcement servers and network access servers.

In order to provide seamless interoperability of call and service control between PSTN and next-generation (packet-based) network domains, the MGC needs to exchange control messages reliably and securely to the SS7 network via the signaling gateway (SG; it can use the SCTP protocol, RFC 2960, as discussed later). Note that in the PSTN network, the call control and signaling intelligence reside in the SS7 network.

MGCP is currently enjoying the widespread approval of cable TV (CATV)-based VoIP service providers (e.g., see PKT-SP-EC-MGCP-I04-011221.pdf at www.packetcable.com/specifications/). Both IETF and ITU-T's study group 9 (Integrated Broadband Cable and Television Networks Study Group) are considering approval of the extensions of MGCP (MGCP v2, RFC 2705-bis, etc.). MGCP is also evolving to ITU-T's H.248 recommendation [6,7] and IETF's Media gateway control protocol (RFCs 3054, 3015, and 2805).

## SWITCH-BASED VERSUS SERVER-BASED VoIP

For switch-based VoIP services, interworking with the existing PSTN switches, networks, and terminals is desirable. In such scenarios, H.225 and H.245 are well-established signaling and media control protocols under the H.323 umbrella protocol. Note that H.323 defines IP-PSTN GWs, call controller or GK, terminal equipment (TE), and multipoint control units (MCUs) as the elements of the system architecture. H.248/Megaco appears to be the most promising emerging protocol that can complement both H.323 and SIP when SIP/H.323 is used for communication between TEs, and between TE and MG or GW.

For server-based VoIP services, the intended network consists of servers and IP routers. In these scenarios, SIP and its many variants are most useful. For large networks, IETF suggests the use of the TRIP (it defines telephony routing over IP in a fashion similar to that of the BGP; RFC 2871, a work in progress in IETF's IPTel WG, RFC 2871) protocol to locate the server to which a call should be routed. For routing a call from an SIP or IP phone to a PSTN terminal (analog or digital POTS phone), one must use the IP-PSTN GW, call controller, and an ENUM server. ENUM (electronic numbering, RFC 2916) converts the E.164 telephony address to an IP address and vice versa using an enhanced domain name system (DNS) server.

## H.225 AND H.245 PROTOCOLS

Although there are a large number of protocols and standards for signaling and control of real-time VoIP calls, ITU-T's H.22x and H.32x recommendations (details are available at www.itu.int/itu-t/) are by far the most widely deployed first-generation VoIP protocols, especially for international VoIP calls. The key network elements for operation of the H.323 protocol are the IP-PSTN media gateway (MG), a call controller or GK, a multipoint control unit (MCU), and TEs. All of these elements are connected to form the zone shown in Figure 3-1, using a LAN where the quality of transmission cannot be controlled.

The H.225 standard defines ITU-T's Q.931 protocol (a variation of ISDN user network interface layer-3 specifications for basic call control) based call setup and RAS (registration, admission/administration, and status) messaging

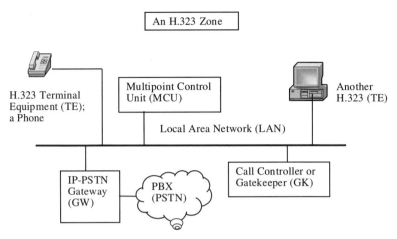

**Figure 3-1**    Network elements and their interconnection using a LAN in an H.323 zone. Note that the PBX (PSTN) is outside the scope of H.323 and is shown to demonstrate the interoperability of H.323 with PSTN.

from a GW or end device/unit or TE to a GK. RAS messages are carried over UDP packets; these contain a number of request/reply (confirmation or reject) messages exchanged between the TE/GW and the GK. TEs can use RAS for discovering a GK or to register/deregister with a GK. A GK uses the RAS messages to monitor the endpoints within a zone and to manage the associated resources.

H.245 defines in-band media and conference control protocols for call parameter exchange and negotiation. These parameters include audiovisual mode and channel, bit rate, data integrity, delay, and so on. They provide a set of control functions for multiparty multimedia conferencing, and can also determine the master/slave relationship between parties to open/close logical channels between the endpoints. In Figure 2-7 I showed the functions and relative positions of H.225 and H.245 with reference to ISO's open system interconnection (OSI) stack [1]. Figure 3-2 shows the protocol sequence for establishment of a real-time H.323 voice communication session from one PSTN phone to another over an IP network. Note that in this diagram, ARQ stands for Admission Request, ACF for Admission Confirm, LRQ for Location Request, and LCF for Location Confirm. Ingress and egress gateways are indicated by IGW and EGW, respectively. Ingress and egress gatekeepers are indicated by IGK and EGK, respectively.

## SESSION INITIATION PROTOCOL (SIP)

SIP (IETF's RFC 3261) refers to a suite of call setup and media mapping protocols for multimedia (including voice) communications over a wide area net-

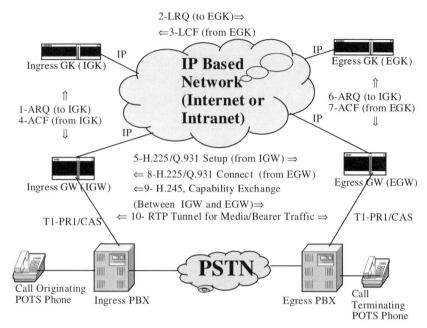

**Figure 3-2** Message exchange for setting up an H.323-based VoIP session from one PSTN phone to another over an IP network.

work (WAN). It includes definitions of the SIP, Session Announcement Protocol (SAP), and Session Description Protocol (SDP; RFCs 3266, 3108, and 2327).

SIP supports flexible addressing. The called party's address can be an e-mail address, a URL, or ITU-T's E.164-based telephone number. It uses a simple request-response protocol with syntax and semantics that are very similar to those of the HTTP protocol used in the World Wide Web (WWW). As the name suggests, SIP is used to initiate a session between users, but it does so in a lightweight fashion. This is because SIP performs location service, call participant management, and call establishment but not resource reservation for the circuit or tunnel that is to be used for transmission of information. These characteristics of SIP appear to be very similar to the features of the H.225 protocol. SAP is used along with SDP to announce the session descriptions proactively (via UDP packets) to the users.

SDP includes information about the media streams, attributes of the receiver's capability, destination address(es) for unicast or multicast, UDP port, payload type, and so on. The receiver's capability may include a list of encoders that the sender can use during a session. These attributes can also be renegotiated dynamically during a session to reduce the probability of congestion. These characteristics of SDP appear to be very similar to the features of the H.245 protocol.

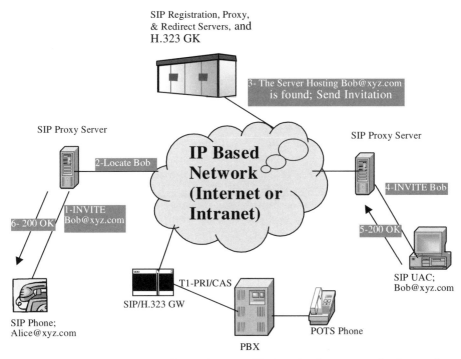

**Figure 3-3**   Message exchange for setting up a SIP-based voice communication session from one IP or SIP phone to another.

SIP architectural elements include (a) user agents (UA): client (UAC) or server (UAS) and (b) network servers: redirection, proxy, or registrar. The client or end device in SIP includes both the client and the server; hence, a call participant (end device) may either generate or receive requests. SIP requests can traverse many proxy servers. Each proxy server may receive a request and then forward it to the next-hop server, which may be another proxy server or the destination UA server. A SIP server may act as a redirect server as well. A redirect server informs the client about the next-hop server so that the client can contact it directly.

Figure 3-3 shows the message exchange for a SIP-based call setup. Note that the number of messages that need to be exchanged to set up a SIP session is smaller than that for an H.323 session (Fig. 3-2). As of 2001, both software-based (running in a PC) and hardware-based SIP and IP phones were available. For call routing over a large IP network, SIP may use the TRIP (telephony routing over IP, a work-in-progress in IETF's IPTel WG, RFC 2871) protocol to locate the server to which a call should be routed. For routing a call to a PSTN terminal (POTS phone), it may be necessary to use the ENUM (electronic numbering, RFC 2916) protocol. ENUM converts E.164 telephony address to IP address (using an enhanced DNS server) and vice versa.

SIP's request-response messages include an INVITE request followed by a reply indicating the results; for example, a reply of 200 OK means that the connection request has been accepted. The request contains header fields that are used to convey call information. Following the header fields is the body of the message, which contains a description of the session to be established.

Since SIP allows the use of fast, (call) stateless proxies in the core of the network and (call) stateful proxies at the edge, SIP is significantly more scalable than H.323. However, one can argue that by using RAS-only GKs in the core and full routing GKs on the edge, it is possible to achieve the same range of scalability in the H.323 domain as well.

The following variants of SIP have emerged during the past few years: SIP-, SIP best common practice (SIP-BCP), SIP+/SIP-T, and so on. However, SIP+/SIP-T appears to be the most useful (feature-rich) variant and the dominant one for interworking with PSTN.

SIP+/SIP-T is an extension of SIP that allows call termination to the PSTN. It encapsulates SS7 ISUP, Q.931 ISDN, or CAS signals as a MIME attachment to a SIP (e-mail) message. SIP+ adds the ability to handle carrier signaling and tunnel PSTN-to-PSTN calls through an IP network. The MIME encoding permits the signals to be tunneled between media gateway controllers (MGCs). SIP+ retains full SIP functionality in the sense that the following features can still be used: (a) multihop searches to route calls to the terminating end, (b) network to network connections (NNIs) to terminate calls to a carrier other than the originating one, and (c) addition of proxies to subdivide the network (which makes it more scalable).

SIP programming interfaces include the call-processing language (CPL), the SIP common gateway interface (SIP-CGI), and SIP server-based applets (servlets).

CPL (RFC 2824) is an extensible markup language (XML)-based scripting language for describing call services. It offers primitives for making decisions based on call properties and is engineered for end-user service creation via graphical user interface (GUI)-based tools. It is fast, lightweight, and scalable.

SIP-CGI is similar to HTTP CGI (almost 90% equivalent). It is the interface that generates SIP services using the programming language of choice. This is very similar to the development of dynamic Web content. It is more flexible than CPL, but doesn't scale as well and can be much more risky to execute. It needs to be guarded against intentional or unintentional malicious script behavior. SIP servlets mirror the concept of HTTP servlets. This is similar to CGI, but the process runs within a JAVA virtual machine (JVM) within servers. The servlets have less overhead than CGI, and their execution is protected within the JAVA "sandbox" construct. The system is more flexible and scalable than CPL.

Because of its simple, flexible, and modular architecture, SIP can be viewed as a simpler, lightweight alternative to the H.225 signaling protocol (used in H.323). Both H.323 and SIP assume RTP for media flows. Since SIP uses the HTTP messaging format and URL for addressing, it can be easily integrated with Web, e-mail, or other existing IP-based services (e.g., instant messaging) and applications. SIP also supports many advanced POTS call features like caller ID, caller name/number mapping services, call waiting, call forwarding, call hold, automatic call distribution, user location and follow-me services, and so on.

SIP appears to be one of the most promising signaling and control protocols for VoIP services. Many vendors are writing SIP software with the objective of executing it on general-purpose computers/servers. Some service providers are working with them to enhance the features of SIP servers so that they can have the functionality of a softswitch. Such a softswitch may perform the functions of a call controller, MGC, and SS7 SG.

SIP and H.323 interworking issues are currently being discussed by the IETF. Both software- and hardware-based solutions and products are being proposed and implemented by the vendors. Technical comparison between SIP and H.323 can be found at the following websites:

a. www.iptel.org/info/trends/sip.html
b. www.cs.columbia.edu/~hgs/sip/h323-comparison.html

## MGCP AND H.248/MEGACO

The media gateway control protocol (MGCP, RFC 2805) is IETF's work-in-progress that is currently being replaced by the ITU-T's H.248/IETF's Megaco protocol. ITU-T's SG 16 developed a competing protocol, H.GCP, but then they agreed to combine their efforts in the Megaco (Media Gateway Control, RFC 3015) standard protocol.

MGCP was created by merging Cisco and Telcordia's (formerly Bellcore) simple gateway control protocol (SGCP) and the media control portion of Level 3's IP device control (IPDC) protocol. MGCP offers a mechanism for decomposing a telephony gateway into a signaling or call control component and a controlled media component, focusing on centralized control of distributed telephony gateways. MGCP assumes a distributed system of IP telephony GWs covering network elements (NEs) like call controllers (CCs), MGCs or call agents (CAs), MGWs, and SGs.

MGWs convert circuit switched (PSTN) traffic into packet domain (IP, ATM, etc.) traffic. They may also perform transcoding functions such as accepting G.711 coded PSTN domain voice traffic and delivering G.729 or G.723 coded voice traffic to the packet domain. Some of the advanced MGWs supports hardware-based echo cancellers, sophisticated packet buffering techniques, and FEC and/or interpolation-based packet voice reconstruction to improve voice quality.

The CC, or MGC, or CA is the device where the call control intelligence resides. In H.323 it is called a *gatekeeper* or *call controllers*, in MGCP a *call agent*, and in Megaco the MGC. Its main function is to provide call control and routing intelligence. It controls several MGWs and SGs.

SGs provide interworking of a packet domain CC (e.g., an H.323 gate-keeper) with PSTN's SS7 network, mainly to interpret call control and service delivery–related messages. It interfaces with the SS7 network using A- or F-type links [1] and with the CC using IP links.

The softswitch concept probably originated during the development of MGCP. A softswitch is sometimes referred to as a collection of software-defined entities residing in general-purpose computers/servers. Theses entities help create, manage, control, and bill telephone calls and related services. Therefore, a collection of servers—hosting the H.323 gatekeepers, SIP servers, MGC, CA, SS7 SG, and so on—could be considered a softswitch.

MGCP can interoperate with H.323 clients, but its main focus is on PSTN-to-PSTN connections via an IP network. An MGCP phone uses CC-based intelligence and features but is incapable of supporting any advanced packet network-based features. Also, unlike SIP users, it cannot place a call without the mediation of the controller. Like MGCP, H.248/Megaco assumes a sepa-ration of signaling (call) control from the MGW. An MGC handles the control function. Since it adds support for media control between TDM and ATM networks and some other flexibility and features, Megaco can be considered a superset of MGCP.

Figure 3-4 shows the salient features of MGCP, and Figure 3-5 presents the prominent characteristics of the Megaco/H.248 protocol.

Currently, H.248/Megaco does not address QoS support issues explicitly

| MGCP: Master ⟶ Slave<br><br>- *Endpoint, Connection*<br><br>- *1 Transaction=1 Command, 1 Type of Response*<br><br>- *Grammar: Text Encoded (BNF)*<br><br>- *Assumes Limited Intelligence in Edge* | Gateway Commands:<br><br>• Notify<br><br>• DeleteConnection<br><br>• RestartInProgress |
|---|---|
| **MGC/Call-Agent Commands:**<br><br>• EndpointConfiguration<br><br>• NotificationRequest<br><br>• CreateConnection<br><br>• ModifyConnection<br><br>• DeleteConnection<br><br>• AuditEndpoint<br><br>• AuditConnection | **Packages Defined in the MGCP Specs:**<br><br>• Generic Media Pkg<br><br>• DTMF Pkg<br><br>• Trunk Pkg/Line Pkg<br><br>• Handset Pkg<br><br>• RTP Pkg<br><br>• Announcement Server Pkg |

**Figure 3-4** Major features of the MGCP protocol. (Source: IETF, RFC 2705, 1999, 2000.)

| MeGaCo: Master→Slave | MGC/Call-Agent Commands: |
|---|---|
| - *Terminations, Context* | • Add |
| - *1 Transaction=N Actions;* | • Modify |
| - *1 Action=N Commands,* | • Subtract |
| - *2 Types of Responses* | • Move |
| - *Grammar: Text Encoded (ABNF) and Binary Form (ASN1)* | • AuditValue |
| - *Assumes limited Intelligence in Edge* | • Notify |
|  | • ServiceChange |

**Figure 3-5** Major features of the Megaco/H.248 protocol. (Source: IETF and ITU-T, RFC 3015, 2000.)

and is not backward compatible with MGCP. In addition, it does not address MGC-to-MGC protocols. It is reasonable to expect that SIP, SCTP, or BICC (discussed later) will be useful in solving these interworking problems.

In addition to supporting SIP, many VoIP-related industry forums and vendors are currently focusing their activities on MGCP and Megaco/H.248. The Multiservice Switching Forum (MSF) and the International Softswitch Consortium (ISC) announced the results of the first Megaco/H.248 interoperability event held at the University of New Hampshire (UNH) Lab. The event included tests of media flow. Although most of the implementations used the real-time protocol (RTP) on an Ethernet network, one of the MG implementations had an ATM network for media transmission as well. Up-to-date information on findings and issues discovered during these interoperability studies are available at the websites of MSF (www.msforum.org), and ISC (www.softswitch.org).

## STREAM CONTROL TRANSMISSION PROTOCOL (SCTP)

SCTP (IETF's RFC 3057 and RFC 2960) is IETF's Signaling Transport (Sig-Tran) work group's newly recommended protocol. SCTP addresses the transport of SS7 signaling messages like ISDN (Q.931), ISUP, and so on between various network elements—such as the SG, MGC, and MGW—over packet-based (IP) networks.

SCTP is a reliable datagram (transport layer) protocol. The adaptation layers have been defined for the transport of TCAP (Transaction Capability Application Part), ISUP (ISDN User Part), MTP-2, and MTP-3 messages. SCTP provides better security, timing, and reliability than the existing TCP/UDP-based transport mechanism.

The primary features of SCTP are (a) backward compatibility with UDP, (b) acknowledged, error-free, and nonduplicated transfer of user data, (c) sup-

port of data segmentation to conform to the discovered path message transmission unit (MTU) size, (d) guaranteed in-sequence delivery of user messages
within multiple streams, (e) multiplexing (optional) of user messages into SCTP
datagrams, and (f) support of multihoming to achieve network-level fault tolerance.

In order to make it conform to SS7 signaling standards requirements, SCTP
has been designed to support stringent packet loss, network delay, and security
requirements. Therefore, it is highly likely that SCTP will be adopted as
another standard signaling transport protocol in IP-based networks.

SCTP uses UDP in the transport layer. In another proposal (IETF's RFC
3094), a protocol called transport adapter layer interface (TALI) has been recommended for adapting MTP3-encapsulated TCAP and ISUP messages for
transmission over TCP/IP. Only a very limited number of SG manufactures are
supporting TALI.

A handful of SG manufactures are announcing products that support the
SCTP protocol. Many of them are also participating in the ongoing SCTP
bake-off and interoperability test events.

## BEARER INDEPENDENT CALL CONTROL (BICC)

ITU-T's study group 11 (SG 11) is working on BICC. BICC handles the setup
of packet bearer connections between many types of bearers, such as PSTN
(via SS7), ATM, and IP. It can handle both forward and backward call setup.
It supports an extended CIC (circuit/carrier identification code) field and could
be used for the connection of multipoint gateways. In the United States, the
ANSI Committee T1S1's Common Channel Signaling (T1S1.3) and Services
Architecture and Control (T1S1.7) working groups are developing the North
American position on BICC. Therefore, BICC is expected to support interworking with both ANSI and ITU-T versions of the SS7 protocol. Three types
of serving nodes have been defined within the scope of BICC for the proposed
call control protocol: (a) an interface serving node that provides an interface to
PSTN or circuit switched networks, (b) a transit serving node that provides call
and bearer transit functionality within a single network, and (c) a GW serving
node that provides call and bearer internetwork GW functionality. BICC provides call control between serving nodes only. Bearer control between serving
nodes is provided by other protocols, as shown in Figure 3-6.

BICC's capability set 1 (CS1) defines the operation of the BICC protocol
with the ATM Adaptation Layer-2 (AAL-2) signaling protocol (ETSI EG 201
849, January 2001). The next version of BICC will have capability set 2 (CS2)
to enable it to be carried over IP using SCTP (described above). ITU-T's
Q.1901 and Q.765.5 are the relevant recommendations on BICC. In May 2001,
ITU-T's study group 11 (SG 11) mandated the development of enhanced SIP-
and SCPT-like features in capability set 3 (CS3) of their BICC protocol. BICC-

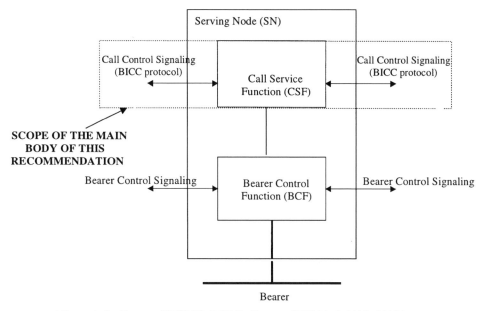

**Figure 3-6**   Scope of ITU-T's BICC. (Source: ITU-T, Q.1901, 2000.)

CS3 may also support network-to-network interfaces (NNI) to H.323 and SIP-based networks. It may take a few years before these standards are completely developed and the vendors start supporting them. As of this writing, only a very limited number of vendors are committed to support BICC.

## FUTURE DIRECTIONS

In this section, the most promising VoIP call control and service-related protocols are identified. Then current industry efforts and protocols for interworking of PSTN and IP domain services are discussed. Finally, an integrated or hybrid VoIP signaling model is presented.

### The Promising Protocols

In the next few years, we expect to see enthusiastic activities in the domain where protocols like SIP, TRIP, ENUM, MGCP, Megaco/H.248, SCTP, and perhaps BICC will be utilized and enhanced for rapid creation, dynamic management, and delivery of next generation services. The network time protocol (NTP, RFC 1305) may also be very useful to synchronize the clocks of various VoIP network elements operating and providing real-time telephony services over a wide geographic area. This synchronization may help improve voice quality and service reliability. NTP can run on a dedicated server, the network time server (NTS), and can distribute timing information from a highly reliable time source (like the global positioning system [GPS] clock) to the routers, servers, and MGWs that are offering the VoIP service.

Because it interworks with PSTN networks, a majority of today's public VoIP traffic is being carried over H.323-based networks, and due to its current level of stability and proliferation, it can now be considered a legacy VoIP protocol. The newer versions of H.323 attempt to incorporate SIP-like features and functions to give H.323 a more Internet-friendly flavor. VoIP customers are also moving rapidly toward SIP-based clients and services. This is because SIP is an edge-centric (intelligent edge) protocol and because Microsoft has announced its support for SIP in the Windows XP operating system.

For switch-based IP centrex services, once again H.323 is considered to be mature enough to satisfy current customer demands. However, here again, the trend is toward using SIP. Hence, the NEs that support graceful interworking and/or migration toward SIP-based services will be the winners.

For server-based IP centrex services, SIP is and will remain the protocol of choice because it encourages innovation at the endpoint or end devices, and follows the Internet paradigm of service creation, distribution, and management very closely. These might have an impact on the demand for T1-PRI links to corporations because most corporations connect their PBXs to the PSTN network using T1-PRI links and sometimes use the PSTN-hosted centrex features via the PBXs.

Many emerging and established carriers are currently carrying tens/hundreds of millions of minutes of VoIP traffic every month[2] to provide basic transport of voice traffic/calls to residential and corporate customers. As this market matures and becomes saturated, customers will demand full-feature voice services irrespective of their location in the enterprise and their ability or willingness to pay.

It appears that the trend in the industry and among vendors is to "SIPize" the H.323 rather than to "H.323ize" the SIP. Even if H.323 is completely SIPized, SIP may be the winner because of its simplicity and its resemblance to the architecture and protocol of the Internet: Its endpoint is an intelligent device (like a PC), and the network is a combination of routers and servers (not switches and mainframes). As mentioned earlier, SIP is roughly equivalent to the H.225 protocol, and the session description protocol (SDP) is generally comparable to the H.245 protocol.

As far as the protocol between MG and MGC is concerned, the use of H.248/Megaco will proliferate, although some may be using H.323 and/or MGCP in this area. The link between SG and MGC can use SIP, although it may be prudent to use SCTP. And for interworking of MGCs, IETF's SIP-T,

---

[2] The Enterprise Networking group of AT&T Business carried over a billion minutes of voice over IP during the year 2000 (Source: NetEconomy, June 11, 2001, p. 22).

SCTP, or ITU-T's BICC-CSx can be used. It remains to be seen whether IETF and ITU-T will merge their activities on SIP/SCTP and BICC-CSx or not. In the past, these two organizations teamed up to develop decomposed call and media control, as well as media translation architecture and protocols. One such example is the H.248/Megaco protocol, which helped to develop further the MGC and softswitch[3]-based architecture for the evolution of PSTN. In general, a better, more stable, and more scalable standard will be developed if these two organizations (IETF and ITU-T) work together to achieve the same goal. We also note that ISC and MSF closely follow whatever ITU-T and IETF jointly develop or recommend.

### Interworking of PSTN and IP Domain Services

IETF's SigTran working group developed SCTP (RFC 2960) for reliable and secure transmission of SS7 messages over IP. IETF's PINT[4] and SPIRITS[5] are currently recommending the use of SIP and SCTP protocols (RFC 2848 uses SIP, and RFC 3055 uses SIP and SCTP) for making telephone call features invokable from IP networks and vice versa. This is because SIP enables rapid creation and cost-effective rollout, maintenance, and delivery of enhanced VoIP connections and services. SCTP is expected to offer robust carriage of PSTN's SS7 signaling messages over IP networks. Figure 3-7 shows the architecture that has been adopted by the joint PINT/SPIRITS working groups of the IETF.

### Hybrid Signaling Model

For a converged network or for a next-generation multiservice network, it may be useful to roll out services based on SIP- and H.248/Megaco-compatible products. In the foreseeable future, we may encounter two legacy networks: the circuit-based PSTN and the H.323-based VoIP network.

In the next-generation multiservice network, an integrated signaling and call control architecture, as shown in Figure 3-8, may be envisioned for enhanced

---

[3] As mentioned earlier, the softswitch concept probably originated during the development of MGCP. A softswitch is sometimes referred to as a collection of software-defined entities residing in general-purpose computers/servers. Theses entities help create, manage, control, and bill the telephone calls and related services. Therefore, a collection of servers—hosting the H.323 GKs, SIP servers, MGC, SS7 SG, and so on—could be considered a softswitch.

International Softswitch Consortuium's website (www.softswitch.org) lists more than 100 small, medium-sized, and large softswitch vendors.

[4] PSTN and Internet Inter-Networking (PINT): PINT uses the IP client and PINT server and allows invoking of telephone call service from the IP domain (see RFCs 3055, 2848, and 2458).

[5] Service in the PSTN/IN Requesting Internet Service (SPIRITS): SPIRITS is developing client-server-based architecture for graceful, secure interaction between intelligent networking or IN (PSTN) triggers and IP domain services. Preliminary services include Internet call waiting, caller ID, and call forwarding.

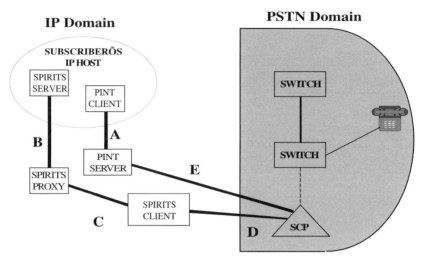

**Figure 3-7** An architecture for making IP domain services seamlessly available to PSTN users and vice versa. This has been adopted by the joint PINT/SPIRITS working groups of the IETF. A, B, C, D, and E are the links where protocols and Interfaces need to be standardized. (Source: IETF, 2001.)

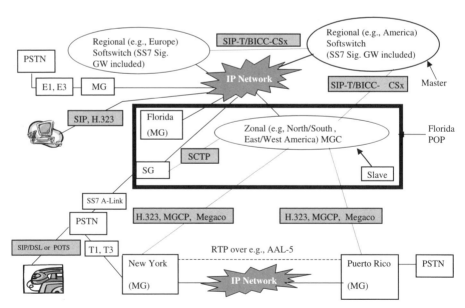

**Figure 3-8** An architecture for a packet-based global network for advanced or enhanced VoIP services.

VoIP services. Figure 3-8 also shows the use of a softswitch in the next-generation multiservice network. This architecture is expected to support SIP/H.323-, MGCP/H.248-, and PSTN-based calls, call routing, and translation of signaling from one signaling domain to another. End users can be connected to the network via POTS phones, SIP user agents, or an H.323 terminal. These end devices can communicate with the call-control complex or softswitch (contains MGC and SG) using the SS7, SIP/H.323, or MGCP/H.248 protocol.

An example of the H.248/Megaco-based call establishment between two communicating entities is shown below. In this case, the MG and MGC use the H.248/Megaco protocol.

- MGC sets the endpoint—which could be an H.323 TE, a SIP UA, an IP/POTS phone, and so on—to look for an *off-hook* indication; for example, the receiver is picked up to make a call.
- The MGW informs the MGC about the off-hook condition. The end systems can use the SS7, SIP, or H.323 protocol to communicate with the MGC.
- The MGW puts a dial tone on the line after receiving the command to do so from the MGC, listens for DTMF tones indicating the dialed number, and passes the number back to the MGC.
- The MGC then performs appropriate routing lookup for the call. It uses an intersoftswitch (or MGC) signaling protocol such as SIP-T or BICC-CSx to contact the call-terminating MGC.
- The call-terminating MGC instructs the appropriate MGW to send a *ring tone* over the line to the terminal device's corresponding dialed number.
- When the MGW detects that the terminal device's corresponding dialed number is off-hook, both MGWs are instructed by their respective master MGCs to establish a two-way tunnel over the packet network. This tunnel could be an RTP tunnel in the case of an IP network, a switched virtual circuit (SVC) over AAL-2, or a virtual circuit over AAL-5 in an ATM network.

## REFERENCES

1. T. Russell, Signaling System #7, Second Edition, McGraw-Hill Book Companies, New York, 1998.
2. H.323 Recommendation, Packet-Based Multimedia Communications Systems, ITU-T, Geneva, 1996, 1998, 2000.
3. H. Liu and P. Mouchtaris, "Voice over IP signaling: H.323 and Beyond," IEEE Communications Magazine, Vol. 38, No. 10, pp. 142–148, October 2000.
4. H. Schulzrinne and J. Rosenberg, "The Session Initiation Protocol: Internet-Centric Signaling," IEEE Communications Magazine, pp. 134–141, October 2000.

5. RFCs: All of the RFCs mentioned throughout this chapter are available at IETF's RFC page at http://www.ietf.org/rfc.html, 2001.

6. H.248 Recommendation, Gateway Control Protocol, ITU-T, Geneva, 2000 (also at http://www.ietf.org/html.charters/megaco-charter.html).

7. T. Taylor, "Megaco/H.248: A New Standard for Media Gateway Control," IEEE Communications Magazine, pp. 124–132, October 2000.

8. H. Schulzrinne, Editor, Converging on Internet Telephony, A Theme Feature of the IEEE Internet Computing, Vol. 3, No. 3, pp. 40–91, May/June 1999.

# 4

# CRITERIA FOR EVALUATING VoIP SERVICE[1]

In this chapter, I describe a set of important criteria that can be used to perform qualitative and quantitative measurements of IP phone or POTS phone (black phone) to black phone/IP phone voice calls over an IP network. Since a legacy POTS call, with all of its robust characteristics over an IP network, is considered to be a killer application (service) by many of the proponents of VoIP, it is recommended that a private IP network or Intranet be used for measuring performance. This is because the network operator has better control over the entire network—ingress, egress, routing paths and protocol, and so on—in such a scenario, and the best possible performance can be achieved when an internal IP network instead of the public Interent is used for VoIP. The performance parameters of interest are availability of the network and dial tone, call setup request processing performance, call completion/drop rate, one-way voice transport delay or voice envelop delay, voice quality during the conversation using both subjective and objective measures, and so on. There is a series (more than 100) of Telcordia LATA switching systems generic requirements (GRs)—commonly known as LSSGR (details can be found at www.SAIC.com, 2001)—which specify the reliability, availability, and service requirements of PSTN switch-based telephony/voice calls. These specifications may need to be revised in the context of VoIP services offered using next-generation packet-switch-based multiservice networks.

---

[1] The ideas and viewpoints presented here belong solely to Bhumip Khasnabish, Massachusetts, USA.

## SERVICE REQUIREMENTS BEFORE CALL SETUP ATTEMPTS

Two of the most important parameters of interest for VoIP even before a call setup attempt is made are the following:

- Availability of the dial tone so that the users get the impression that the call-processing host or switch is *ready* to deliver the service and
- Availability of computing and network resources for honoring call processing requests. This includes collecting information on the called party's identification (e.g., the E.164-based telephone number, e-mail address, URI/URL), processing this information to determine the *best* possible route to set up an RTP/UDP/IP session, and finally, connecting the called party's phone to the calling party's phone.

The traditional PSTN networks have been designed to provide lifeline services such as processing of emergency or 911 calls. Therefore, in the United States, PSTN service providers must design their networks to deliver the dial tone to the customer's phone 0.30 to 3 sec after the handset is picked up in 95% of instances. This must happen even when the electric power supply is not available. If VoIP is used for voice transmission service only, it may not be difficult to satisfy this requirement, because the dial tone will still be delivered from a PSTN switch. However, if VoIP is to be used ETE, including the customer's premise equipment—if, for example, an IP phone is used at home instead of a POTS phone—the access routers and call processing servers must be designed to satisfy the above-mentioned stringent availability requirements unless the regulations are relaxed for IP-based real-time voice telephony services.

Next, it is well known [1] that the PSTN network has been designed with low utilization of transmission and processing resources in mind. That is, the probability that all of the users who are connected to one PSTN switch will pick up the phone to make a call at the same time is very low. This may not be the case for IP networks/protocols that are being experimented with and redesigned to support data, voice, and video services. Therefore, when multiple data- and graphic-sharing sessions are in progress in an IP network, the edge devices and network may not have enough resources to honor a call processing request unless a certain amount of these resources are reserved for processing VoIP calls. This requires the operation of an IP network in overprovisioned or service-based resource allocation mode, which may not be very cost-effective, although it is practically achievable.

## SERVICE REQUIREMENTS DURING CALL SETUP ATTEMPTS

One of the most important requirements during a VoIP call setup attempt is the call processing performance, which includes the following two factors:

- The total amount of time it takes to set up a call, measured from the moment the last digit of the first-stage dial-in number—as in multistage dialing—is entered to the moment the ring-back tone is heard at the call-originating side. In IP telephony, call setup time can vary from 500 msec to 10 sec, depending on the availability of network and digital signal processing (DSP) resources in the system being used. This refers to the call setup time in an idle system. I discuss these and related issues in Appendix A.
- The number of simultaneous calls that can be handled without any precall wait. This refers to setting up a call in a busy system. Note that the precall wait can vary from as little as 1 sec to as much as 10 sec, depending on the speed of the CPU used in the IP-PSTN GW, availability of memory/storage and (digital signal) processing resources in the system, and so on. I discuss these and related issues in Appendix B.

In addition, there may be requirements to support network-level prioritization of calls, depending on the number from which the call is originating or the number for which the call is destined.

It is widely believed that because of sharing of resources and the routed (instead of switched) nature of connections in operational VoIP networks, the call processing performance will be, at most, as good as it is in cellular or wireless networks. In PSTN networks, regional and national call setup time may vary from ~2 to 4 sec (see, e.g., the section on call setup time at www.att.com/network/standrd.html, 2001), depending on whether or not database lookup is needed. Note that database lookup is required for credit card–based calls, toll-free calls, and other types of calls.

According to ITU-T's E.721 recommendation [2], the average answer-signal delay (the delay between the time the called party picks up the receiver and the time the caller receives an indication of this) should be 750 msec for local calls, 1.5 sec for toll calls, and 2.0 sec for international calls, with 1.5, 3.0, and 5.0 sec as the 95% values, respectively. ITU-T's E.721 recommendation [2] also states that the average postdial delay (the interval between dialing the last digit and hearing the ring-back tone) should be no more than 3 sec for local calls, 5 sec for toll calls, and 8 sec for international calls, with 95% values of 6, 8, and 11 sec, respectively.

To deliver PSTN-grade call processing performance, the edge devices, servers, and IP network itself must be designed to be as robust and have as high a capacity as the PSTN system. This may not yet be very cost-effective to implement.

## SERVICE REQUIREMENTS DURING A VoIP SESSION

After a VoIP session is established, the packetized voice signal must be delivered from the source (talker) to the destination (listener) in real time without compromising the integrity of the signal. The relevant parameters of interest are voice coding, processing, envelop delay, packet loss, voice frame packing, bufferring, reconstruction (e.g., delay jittering) strategies, and so on, as discussed below. The situations become more challenging when one attempts to make

a. PSTN-hosted advanced services and call features—such as the caller's name and identification (ID), call waiting, and three-way call—available to IP domain clients like PCs and IP phones, and/or

b. IP domain features or Internet-hosted services—such as unified messaging, buddy list and follow-me services, and media conversion and sharing—available to analog/digital or ISDN phones.

In addition, there is a series of standards (in PSTN) for echo cancellation, billing, network- and service-level testing and diagnosis, and regulatory function (e.g., identifying the caller's location for 911 calls, call tracing and recording for supporting CALEA, etc.) related requirements. These can be found in various ITU-T standards documents and in Telcordia's (www.saic.com/about/companies/telcordia.html, 2001) LSSGRs.

**Voice Coding and Processing Delay**

The voice coding and processing delay consists of the delay incurred due to (a) analog to digital conversion, (b) packetization or framing, (c) packing of frames, (d) incorporation of error-correction mechanisms, loss- and privacy-protection mechanisms, and so on of the voice signal at the sender's end. These processes are executed in reverse at the receiver's end, and a similar delay is incurred there too. These delays are shown in Figure 2-1.

Many of the newly developed low-bit-rate voice coding schemes like ITU-T's standards G.723, G.729, and so on are now commonly utilized for VoIP applications. These schemes utilize advanced memory (or buffer) management and digital signal processing (DSP) techniques to generate low-bit-rate voice streams, and hence may add significant coding and processing delay. For example, as discussed in Chapter 2, the coding delay for G.723.1 ACELP (5.3 Kbps) and G.729 CS-ACELP (8 Kbps) schemes could be as high as 37.5 and 15 msec, respectively, in comparison with zero coding delay for the G.711 PCM (64 Kbps) coding scheme. Further delay would be incurred when additional error-correction and loss- and privacy-protection mechanisms are utilized. As a general rule, for G.711 coding at either the sending or the receiving network, the coding and all processing delay should not exceed 15% of the overall mouth-to-ear (M2E) delay. The M2E delay (discussed below) value recommended by the ITU-T in the G.114 specifications [3] is 150 ms if one wishes to maintain the toll quality (MOS value of 4.0) of voice. Thus, for G.711 coding, for ETE VoIP service when the calls are made from one IP phone to another, the total delay in the access or delivery network should not exceed 22.5 msec (i.e., 15% of 150 msec). This leaves 105 msec as the maximum allowable delay (tight upper bound) in the transport or backbone network.

When advanced coding mechanisms (e.g., G.723, G.729) are utilized, the delay incurred in the receiving or sending network could be as high as 30% of the 150 msec, and the delay budget for the transport network is reduced to as little as 60 msec. These scenarios call for deployment of very-high-speed links in the transport network and operating them at very low short-term utilization rates.

### Voice Envelop Delay

Voice envelop delay is the ETE one-way voice transport delay. The delay—commonly known as *M2E delay*—is measured from the moment a noticeable voice signal appears at the sending end (speaker's mouth) of a connection to the moment the same voice signal appears at the receiving end (listener's ear) over an established connection. It includes the voice signal framing, packetization, and buffering delays at the sending and receiving ends, as well as one-way network transport (signal propagation and transmission, packet switching, routing and queueing, etc.) delay.

As shown in Figure 2-1, the one-way network transport delay consists of (a) switching, routing, and queueing delay at the ingress (access) and egress (delivery) networks and (b) transport network or transmission delay including signal propagation delay. As mentioned in the previous section, the general rule is to keep the one-way transport (or backbone) network delay below 70% (for G.711 coding) of the overall M2E delay (150 msec) recommended by the ITU-T's G.114 specification [3] if one wishes to maintain the toll quality (MOS value of 4.0) of voice.

Usually, the ingress and egress network packet transfer delay values are significantly less than those in the transport network. This is due to the fact that it is easy and relatively inexpensive to overengineer the ingress and egress networks in order to operate them in overprovisioned mode. The transport network delay is predictable in switched networks like PSTN and ATM networks, but IP networks like the Internet are routed networks, and they support transmission of a variety of real-time and non-real-time traffic over the same network. Consequently, packet queueing and routing delay contribute significantly to transport network delay even when higher-speed links are deployed, as discussed in Chapter 2. For example, the time required for transmitting a 128 byte (or a 7 msec sample of G.711 or PCM, encoded voice, as shown in Fig. 2-2) VoIP packet over an idle or lightly utilized 128 Kbps WAN IP link is $[(128 \times 8)/(128 \times 10^3)]$ or 8 ms. This delay value can become 15 msec when the link becomes moderately ($\sim 40\%$) utilized and 50 msec when the link becomes heavily ($\sim 90\%$) utilized. This is due to the fact that the queues (at both the ingress and egress of a link) build up very quickly as link utilization increases. To alleviate this problem, any one or more of the following techniques can be used: (a) reduce the size of the VoIP packets by using a smaller voice sample and/or compressing the RTP/UDP/IP headers by using the recommendation suggested in IETF's RFC 2508; (b) use IETF's IntServ, DiffServ (TOS byte setting, as suggested in IETF's RFC 3246 and RFC 3247), MPLS tagging, etc.) to offer higher emission priority to voice packets; (c) use multiple moderate-speed WAN IP links for VoIP applications, and periodically monitor the utilization of these links in order to route the VoIP calls and traffic in real time through the least utilized link; and (d) use higher-capacity links when both real-time voice and non-real-time data traffic are transmitted over the same link.

The signal propagation delay depends on (a) the physical distance between the talking party or sender and the listener or receiving party and (b) whether electrical or optical signal transmission is used. Electrical signal propagation delay varies from 8 to 10 msec per 1000 wire miles, depending on the quality of the materials used to make the wire (or the medium). Optical signals travel faster than electrical signals, and hence it is recommended that high-quality fiberoptic links be used for very-long-haul (tens of thousands of wire miles) transmission of real-time packetized voice over IP networks.

To measure this M2E delay, we used an oscilloscope-based setup, as described in Chapter 5. The probes of the oscilloscope are connected to the outgoing monitor channel of the Hammer tester (described in Chapter 5 and in the appendixes) and to the incoming lines of a BRI phone. The time axis (horizontal axis) of the scale is set to measure the values with millisecond resolution. Using this setup, it is possible to make a call from the Hammer tester (using Hammer script) to the BRI phone and play a voice prompt (e.g., "Hello") at a prespecified time interval. The outgoing and incoming signals can now be monitored in the oscilloscope, and the time difference can be measured. This procedure gives an estimated measure of the ETE one-way voice transport delay. This delay can be measured with no background connections or with a prespecified number of connections or conversations in progress in the background. Excessive one-way voice transport delay is quite common in packet-switched networks. This impairs the quality of the voice by adding echo, and could be very annoying to the calling and called parties.

As mentioned earlier, ITU-T's G.114 standard [3] states that the one-way ETE voice transport delay should be below 150 msec for good-quality real-time speech communications. ITU-T's G.131 standard [4] provides specifications for talker echo control. It states that the degree of user-perceived annoyance depends not only on the extent to which the echo from the original speech is delayed, but also on the difference in amplitude between the two. Usually, the echo must be suppressed or cancelled when the one-way delay from talker to listener exceeds 25 msec. ITU-T's G.168 standard [5] specifies methods and ranges (tail length of 128 msec) for the operation of digital line echo cancellers. Echo cancellers use special-purpose DSP chipsets to monitor the incoming speech samples from the talker at the *far end* and produce a delayed estimate of the electrical echo resulting from the *near-end* reflections. This estimate is then subtracted from the composite speech—mainly composed of the *near-end* talker's voice—destined for transmission back to the *far end*.

**Voice Packet Loss**

VoIP uses UDP to transport RTP-encapsulated voice frames. Therefore, in an IP network (Intranet or public Internet), which is supporting transmission of both TCP and UDP streams, loss of voice packets happens when buffers overflow, due either to excessive amount of traffic in the network or to a large number of TCP streams in the network. In addition, corruption of bits during transmission may lead to loss of UDP packets. Loss of voice packets is not only irritating to the communicating parties, it may also give the impression that the call has been dropped. Depending on the packet transmission delay and network delay jitter, a packet loss of up to 10% may still produce acceptable quality (MOS value of 3.0 to 4.0) of voice signal [6]. Common methods for concealing the effects of voice frame loss from the listener's ear are the following: (a) silence or comfort noise can be played in place of the lost frame(s); (b) the latest good speech segment can be repeated; and (c) speech synthesis, repetition, interpolation, and code book (vocabulary book) techniques can be utilized to reproduce the lost frame(s).

Note that since the RTP header includes the packet sequence number, it is possible to calculate the packet loss ratio as the number of lost or missing packets to the total number of packets transmitted during a conversation. Many researchers have also proposed bit, byte, voice frame, and packet-level interleaving and forward error correction (FEC) methods to reduce the impact of packet loss in voice over IP applications.

Voice frame interleaving can be used to reduce the effect of packet loss on voice transmission. At the transmitting end, a voice frame can be divided into multiple segments, and each of these segments can be transmitted over non-adjacent packets in a packet voice stream. At the receiving end, the retrieved segments of a voice frame need to be rearranged in their original sequence. In case of sporadic packet loss in the network, this practice results in one or more short glitches/fractures in the received voice stream. The only drawback of the frame interleaving technique is that it increases the voice signal reconstruction time.

FEC implementation within RTP to protect both the RTP header and speech has also been suggested [7]. For example, half of the previous voice frame and half of the next voice frame can be added to the current voice frame before encapsulating it using the RTP header. This strategy definitely increases the packet size, and hence causes greater delay and needs higher bandwidth for voice transmission, but it may be helpful in reconstructing an erroneous voice frame at the receiver's end in a timely fashion. This would certainly improve the quality of the received voice signal.

**Voice Frame Unpacking and Packet Delay Jitter Buffer**

For VoIP, the addition of RTP, UDP, IP, and Ethernet/PPP headers makes the amount of overhead per voice frame too large (e.g., see Fig. 2-2). To overcome this problem, header compression can be used or multiple voice frames can be packed together before encapsulating them into one RTP packet. A trade-off exists between the number of voice frames that can be packed into one packet and the amount of delay due to large packet size, which may cause degradation of voice quality.

As mentioned in Chapter 2, the packet delay jitter buffer at the receiver's end attempts to neutralize the effects of variation in delay from one packet to the next in the incoming packet voice streams (RTP streams). An artificial delay is added to each packet's arrival time in order to write asynchronously the incoming packets at the head of the buffer. The speech frames are extracted from the tail of the same buffer at a steady rate for proper playback. This delay jitter buffer is elastic, and its occupancy level is allowed to grow and shrink to accommodate the delay variations. However, the delay jitter buffer should never be allowed to underflow or overflow. This may cause breaks or discontinuity in the reproduced voice signal. Belated packets may cause the buffer to underflow, and precocious packets may cause it to overflow. This can be avoided by setting a threshold in the delay jitter buffer and delaying the playout until the buffer occupancy exceeds this threshold. This threshold value must be dynamic enough to reflect the changing network conditions. As a result, the additional delay due to this buffer would not adversely affect the voice quality.

Depending on the delay budget, type of CODEC, voice sample/packet size, and implementation complexity, the delay jitter buffer's size may vary from two to four voice samples/packets. This is equivalent to 40 to 80 msec of delay for a voice sample/packet size of 20 msec, for example, in a G.711 or 64 Kbps PCM-coded voice signal. For G.729 (CS-ACELP, 8 Kbps) coding 15 msec of coding delay is incurred, and if 30 msec of voice sample/packet size is used, the delay jitter buffer size may become 60 msec or higher. The situation gets worse when G.723.1 (ACELP, 5.3 Kbps) coding is sued, because a coding delay of 37.5 msec is incurred. A larger delay jitter buffer may cause degradation in voice quality unless appropriate hardware-based echo cancellation is deployed.

**Management of Voice Quality During a VoIP Session**

In circuit switching, once a circuit is allocated for a voice call or connection, the quality of transmission of the voice signal is almost guaranteed for the duration of the conversation. This is not the case in packet-switched networks unless an emulated circuit—for example, the ATM technology that supports circuit emulation service [8] for real-time applications like real-time voice conversation—is allocated for the service.

As suggested in Chapter 2 and mentioned earlier, any one or more of the techniques discussed in the following paragraphs can be used to maintain the desired quality of voice transmission during a VoIP session.

The size of the VoIP packets can be reduced by using a smaller voice sample and/or by compressing the RTP/UDP/IP headers by using the recommendation suggested in IETF's RFC 2508. This may call for more processing, buffering, and bugger management delay at both the sender and receiver sides. These delay budgets must be carefully managed in order to maintain an acceptable level of voice quality.

IETF's IntServ (RSVP signaling), DiffServ (TOS byte setting, as suggested in IETF's RFC 3246 and RFC 3247), MPLS tagging, and so on) can be used along with higher emission priority (using, e.g., IEEE 802.1p/Q at the link layer or layer-2, as shown in Fig. 2-10) to offer transmission precedence to voice packets. These may call for upgrading the software and/or hardware of the routers and switches in the IP network that is being used for transmitting the packetized real-time voice. This may be time-consuming, expensive, or both except in private IP networks. However, in reality, these upgrades will ultimately happen in the public Internet if it is to support delivery of real- and non-real-time multimedia traffic to users.

Multiple moderate-speed WAN IP links can be deployed for VoIP applications, and the utilization of these links can be monitored periodically in order to route VoIP calls and traffic in real time through the least-utilized link. Monitoring of the links can be passive, active, or both, depending on a number of criteria. The amount of software, hardware (databases, server firms, etc.), and additional (overhead) network traffic—for example, for executing the "ping" and "traceroute" commands—would be significantly different, depending on the mechanism deployed. Proactive monitoring of the link's status may include not only monitoring the round-trip delay over a link or measuring its utilization, but also measuring packet delay, delay jitter, packet loss, voice quality, and so on for pilot calls (these are discussed further in Chapter 8). The information obtained from this practice can be used for network capacity planning, engineering, and customer retention as well.

Higher-capacity links can be deployed when both real-time voice and non-real-time data traffic are transmitted over the same link. In general, this depends on the availability of budgets and facilities. Since newer, cost-effective technologies (DSL, cable modem, gigabit Ethernet, Ethernet in the first mile, etc.) are becoming available, we hope that the carriers and service providers will soon upgrade their Internet service facilities. This may lead to general availability of high-bandwidth WAN IP links at relatively low rates within 3 to 5 years.

## SERVICE REQUIREMENTS AFTER A VoIP SESSION IS COMPLETE

After a VoIP session is complete, the call log and a call detail record (CDR) must be maintained. The CDR must capture and store correctly the anatomy of the call. This helps to analyze what happens before, during, and after the call has ended for billing, testing and diagnosis, network capacity planing and traffic engineering, and other purposes. In some cases, it may be necessary to capture the CDR data in a prespecified format for remote storage using standard protocols in the billing system or server for settlement of prices for regulatory purposes. Additionally, there are requirements to (a) trace the physical location of the caller for emergency or 911 calls even when the caller hangs up, (b) record a conversation to assist law enforcement agencies, and so on. Some standards to resolve these issues are either being discussed or are emerging from organizations like ITU-T, IETF, and the International Softswitch Consortium [9,10].

## REFERENCES

1. R. J. Bates and D. W. Gregory, Voice and Data Communications Handbook, McGraw-Hill Book Companies, New York, 1998.
2. E.721 Recommendation, Network Grade of Service Parameters and Target Values for Circuit-Switched Services in the Evolving ISDN, ITU-T, Geneva, May 1999.
3. G.114 Recommendation, One-Way Transmission Time, ITU-T Geneva, 1996.
4. G.131 Recommendation, Control of Talker Echo, ITU-T, Geneva, 1996.
5. G.168 Recommendation, Digital Network Echo Cancellers, ITU-T, Geneva, 1997.
6. IEEE Network Magazine, IEEE Press/Publishers, New York, Vol. 12, No. 1, January/February 1998.
7. J. Rosenberg and H. Schulzrinne, "An RTP Payload Format for Generic Forward Error Correction," Bell Laboratories and Columbia University: Internet Draft draft-ietf-avt-fec-04.txt, 1998.
8. ATM Forum, ATM Traffic Management Specifications Ver. 4.0, 1996.
9. The Internet Telephony Magazine, 2001 (www.itmag.com, 2001).
10. International Softswitch Consortium (ISC), detailed and up-to-date information is available at www.softswitch.org (2001).

# 5

# A TESTBED FOR EVALUATING VoIP SERVICE[1]

A new service must be prototyped and tested in a laboratory environment before massive deployment. This allows objective and subjective evaluation of the service in question. In addition, the findings can be used for tuning the network operations and performance control parameters, as required for maintaining acceptable QoS (as discussed in Chapter 4).

The testbed presented in this chapter consists of a variety of PSTN and IP domain network elements [1]. These elements are required to emulate PSTN and IP networks, IP network impairments, and elements of SS7 networks like SCP and STP. Other network elements include (a) the network timing server, (b) software- and hardware-based IP and SIP phones, (c) analog and digital (including ISDN BRI) circuit or PSTN phones, and (d) test equipment to emulate and analyze single and bulk phone calls. This testbed is used for a variety of VoIP tests and measurements, as described in Appendixes A, B, and C.

Appendix A discusses how this testbed can be used to measure call progress time in IP telephony. A multistage call setup method is proposed, and its implementation using a set of scripts written in Hammer visual basic (HVB) language (www.hammer.com, www.empirix.com, 2001) is described.

Appendix B presents techniques to determine the bulk-call-setup request-handling performance of IP-PSTN GWs. To achieve this, both call burst size and intercall burst time gap must be determined so that the call setup requests are properly processed. These are implemented using HVB language for testing some commercially available IP telephony GWs.

---

[1] The ideas and viewpoints presented here belong solely to Bhumip Khasnabish, Massachusetts, USA.

**59**

Finally, Appendix C shows how this testbed can be utilized to evaluate the impact of various IP network impairments—such as delay jitter, packet loss, and bandwidth constraints—on voice quality and transmission of DTMF messages over an IP network. HVB language is used to implements the test scripts.

A brief description of the testbed is presented in the next section, followed by a detailed discussion of each of its major components. The test and measurements procedures, and associated HVB scripts, are available in the respective appendixes.

## DESCRIPTION OF THE TESTBED/NETWORK CONFIGURATION

This section presents a high-level description of the network configuration used in the testbed. The interconnection diagram is presented first, followed by a brief description of the functionality of the major network elements of the testbed.

As mentioned earlier, the testbed presented in this chapter consists of a variety of PSTN and IP domain network elements. These are the elements needed to emulate PSTN and IP networks, IP network impairments, and elements of SS7 networks like SCP and STP. Other network elements include (a) the network timing server, (b) software- and hardware-based IP and SIP phones, (c) analog and digital (including ISDN BRI) circuit or PSTN phones, and (d) test equipment to emulate and analyze single and bulk phone calls.

For emulating a PSTN network, any commercially available PBX that can support multiple TI CAS/PRI lines and multiple types (analog, digital, ISDN BRI, etc.) of phones can be used. However, in order to support T1- and/or DS3-type intermachine trunks (IMTs), it may be necessary to use a captive CLASS-5 switch like Lucent's (www.lucent.com, 2001) 5ESS switch, Nortel's (www.nortelnetworks.com, 2001) DMS switch, AG Communication Systems' (www.agcs.com, 2001) GTD-5 switch, and so on. An ISDN PBX from Madge (www.madge.com, 2001) called Madge Access Switch 60 and a GTD-5 switch from AG Communication Systems are used in the testbed.

An IP network can be emulated by using multiple EtherSwitches connected via a router that can be programmed to introduce various types of network impairments. For example, NIST-Net (http://snad.ncsl.nist.gov/itg/nistnet/, 2001) and Shunra's (www.shunra.com, 2001) cloud or storm product can be used to introduce IP-layer impairments in a controlled fashion. We use NIST-Net in the testbed described in this chapter.

To emulate the elements of SS7 network elements like SCP and STP, various types of equipment can be used. These include Tekelec's (www.tekelec.com, 2001) MGTS, Eagle's signal transfer point (STP), Inet's (www.inetinc.com, 2001) test equipment, and so on. We use a small STP from Tekelec in our testbed to emulate SS7 network elements, and both MGTS and Inet's spectrum as SS7 test equipment.

Any general-purpose server running the network time protocols (NTPs) (IETF's RFC 1305/1119, RFC 2030, RFC 867/8, etc.) can be used as the IP domain network time server (NTS). For example, TrueTime's (www.truetime. net, 2001, www.truetime.com, 2001) NTS can be used in the testbed. Without proper synchronization of the asynchronously operating IP-PSTN GWs and other IP domain network elements like routers and SIP or IP phones, voice quality and service reliability cannot be assured.

The traditional PSTN switch suppliers such as Lucent (www.lucent.com, 2001), Nortel (www.nortelnetworks.com, 2001), and Siemens (www.siemens. com, 2001) manufacture ISDN BRI and analog and digital phones. We use the BRI phones from Lucent and Siemens, and digital and analog phones from Nortel. IP and SIP phones from a number of suppliers including Pingtel, Siemens, Cisco, Ploycom, and 3Com (e.g., www.pingtel.com, www.siemens.com, www.cisco.com, www.ploycom.com, www.3com.com, 2001) can be used in the testbed described in this chapter.

For emulating PSTN and IP-based telephone calls for tests and measurements, any one or more of the following types of test equipment can be used: Radvision's (www.radvision.com, 2001) test equipment, Hammer's (www. hammer.com or www.empirix.com, 2001) IT, Agilent's (www.agilent.com, 2001) VQT, Spirent's (www.spirentcom.com, 2001) Abacus test system, Ameritec's (www.ameritec.com, 2001) call generation products such as Crescendo/ Niagara, Catapult's (www.catapult.com, 2001) DCT2000, IPNetFusion's EAST product (www.ipnetfusion.com/east.htm, 2001), and Inet's spectrum. Hammer's IT, IPNetFusion's EAST, and Inet's spectrum testers are used in the testbed presented in this chapter.

The network configuration diagram of the testbed is shown in Figure 5-1. The Hammer tester is used for generating bulk emulated PSTN or circuit domain phone calls and for analyzing emulated black (or PSTN) phone to black phone calls. This includes measuring the answer time, the response time at various stages of call progress, and the time required to hear the ring-back tone at the call-originating side. The version of the Hammer tester used in our lab can support a maximum of six T1 lines to the Madge Access Switch.

The analog and ISDN BRI phones can be used to verify the essentials of call progress and to measure audio quality via human perception. Call progress verification includes hearing the generation of appropriate tones—such as a string of DMTF digits, dial tone, and ring-back tone—or a play-out of an appropriate interactive voice response (IVR) message by a human listener.

The Madge Access Switch 60 is a small ISDN PBX or a CLASS-6 PSTN central office (CO) switch. It provides one or more T1-CAS or T1-PRI connections to the PSTN side interfaces(s) of the IP-PSTN GWs under test. In addition, a set of ISDN BRI phones can be directly connected to it. Currently, it has two 8-port BRI cards and several ports to support T1 connections. The BRI cards support eight BRI phones (ISDN 8510T) from Lucent, a set of fax machines and analog phones through Diva ISDN modems, and two BRI

UNIVERSITY OF HERTFORDSHIRE LRC

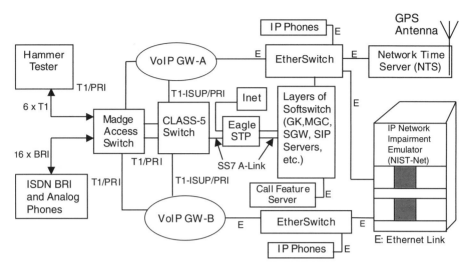

**Figure 5-1**  Block diagram of a VoIP testbed. The softswitch contains various VoIP CC functions, such as, H.323 GK, MGCP/H.248 MGC, and SIP servers for registration, redirect and proxy functions, and may contain the SG and others. The SG can be implemented in a physically separate network element (NE) as well. Clustering or hierarchical interconnections can be used to interconnect the layers of the softswitch. The global positioning system (GPS) antenna attached to the NTS extracts the clock information from a globally synchronized time source and delivers the timing information to all of the IP-based network elements. The IP phones are most likely to be SIP phones.

phones (optiSet NI-1200S) from Siemens. Any of Lucent's BRI phones can support up to 10 calls or connections.

The two 24-port EtherSwitches and the IP network impairment emulator, a PC-based simple router, comprise the Intranet of the testbed. The Ether-Switches provide connectivity to the IP side interfaces of the IP-PSTN GWs (or VoIP GWs) under test.

The VoIP gateway A (GW-A) and gateway B (GW-B) are the near-end (or call-originating) and far-end (call-terminating) GWs. Usually GW-A and GW-B are connected to two different subnets, which are interconnected via the simple PC-based router mentioned above. However, if necessary, it is also possible to connect the two GWs using the same subnet as well, that is, to connect both GWs to the same EtherSwitch. Depending on the type of link interface supported on the PSTN side, an IP-PSTN GW (GW-A, GW-B, etc.) could be either a line-side, trunk-side, or residential GW.

Line-side GWs usually support multiple T1 (CAS or PRI) lines for connectivity to the PSTN network. Trunk-side GWs usually support multiple T1- and/or T3-type intermachine trunks (IMTs) for connectivity to the PSTN network. Residential GWs usually support one (rarely more than one) T1 or digital subscriber line (DSL) line–based connectivity to a CLASS-5 central office switch.

The capabilities of the IP-PSTN GWs are continuously evolving, since the standardization committees and manufacturers are trying to make these devices at least as reliable, available, and capable as the corresponding devices in the PSTN networks. In addition to the IETF and ITU-T websites (www.ietf.org, www.itu.int, 2001), one can find up-to-date information on these devices at the following websites: www.msforum.org, www.softswitch.org, www.pulver.com/von.com, and www.itmag.com.

In general, the softswitch element, which contains the H.323 GK and other call control–related functions, performs registration, administration/authentication, and status (RAS) monitoring functions when a call establishment request arrives. If implemented, it can also maintain the call detail record (CDR) files. A scaled-down version of a softswitch supporting H.323 GK functions can run on a WindowsNT server and can be connected to the same subnet to which GW-A is connected. Note that the softswitch can be considered a more sophisticated version of the GK. It performs all of the required GK functions; supports H.323-v.x GWs, Internet protocol device control (IPDC; see www.l3.com for details), SIP servers, MGCP, Megaco/H.248 devices, and their interworking; and may also support, directly or indirectly, the functions of an SS7 SG. The SS7 SG is a device (server) that provides only a signaling interworking function between the SS7 [2] network and the call controller (CC) functional block defined above. In October 2000, IETF's signaling transmission (SigTran) working group released the stream control transmission protocol (SCTP, RFC 2960) for reliable and secure transmission of PSTN signaling and transaction (SS7 messages like ISUP and TCAP) over IP. Work is also in progress to support adaptation of SS7 MTP level 3 and MTP level 2 messages (M3UA and M2UA) for transmission over IP.

The Inet SS7 tester (www.inetinc.com, 2001) supports a variety of interfaces including V.35, BRI, RS-449, DS0, and DS1 for connections to an SS7 network. It can emulate the SS7 signal transfer point (STP) and service switching and control points (SSP and SCP). Inet can be used to monitor the flow of SS7 messages for a preset group of originating point codes (OPCs) and destination point codes (DPCs). It can be also used to generate SS7 ISUP messages for setting up and terminating PSTN calls, either repetitively or in bulk.

## PSTN EMULATION

For emulating a CO switch of the PSTN network, we use the Madge Access Switch 60 (www.madge.com, 2001) and a CLASS-5 switch such as a GTD-5 switch (see www.agcs.com for details). The Madge switch can accommodate a maximum of six 4- or 8-port cards, with 4 ports in one card reserved for local/remote configuration, network, and timing management. The remaining ports can be used for BRI and/or T1 (CAS or PRI) connections. Currently we are using two 8-port cards for connections to BRI phones and the remaining ports to support T1 connections. The six T1-CAS lines are used to connect to the AG-T1 cards of the Hammer tester, and the remaining T1 lines are used to connect to one or more sets/pairs of GWs under test.

Appropriate dialing plans and Madge switch configurations are used to make connections from one Hammer channel or BRI phone to another, either through the Madge switch directly or via one or two VoIP gateways. These options provide the ability to make calls over the PSTN network/switch alone or through the IP network with incorporation of very little (i.e., when the same subnet is used for connecting the GWs) or a controlled amount of impairments like delay, delay jitter, packet loss, bandwidth restrictions, and so on. These impairments are added in a controlled fashion using an IP network impairment emulator called NIST-Net, as described in detail in the next section.

## IP NETWORK AND EMULATION OF NETWORK IMPAIRMENTS

As mentioned before, we used NIST-Net in our testbed to emulate IP network impairments, that is, to introduce various types of IP-layer impairments in a controlled manner. The IP network used in the testbed is an Intranet. It consists of two Ethernet switches representing two subnets (162 and 146 subnets, i.e., IP addresses 132.197.162.xxx and 132.197.146.xxx are used for the devices connected to the subnets) and a RedHat Linux operating system–based IP network impairment emulator running on a PC. The impairment emulator uses two 10/100 BT Ethernet cards and software from the NIST called NIST-Net (http://snad.ncsl.nist.gov/itg/nistnet/). Basically, NIST-Net is a kernel module expansion of Linux that also offers an X-window-based user interface. It allows addition of a predetermined amount of network impairments such as delay, delay jitter, packet loss, and so on to network performance–sensitive applications in laboratory environments. Since it operates at the IP level, NIST-Net can emulate the critical ETE performance characteristics/dynamics imposed by various WAN impairments (e.g., delay, delay jitter, packet loss, bandwidth restrictions). While characterizing the NIST-Net, we found that the delay and delay jitter values added to the IP streams *do not* exactly *match* the *parameters of distribution* entered in the user interface. Therefore, some modifications[2] were made to the random number generator used in the NIST-Net. Note that we used the ping command to monitor the delay and to calculate the delay jitter added to the IP stream. In addition, an option to add zero delay and a fixed amount of delay alternatively (i.e., zigzag delay) can also be activated when needed. The addition of zigzag delay is helpful in conducting experiments to determine the size of the delay jitter buffer in the IP-PSTN gateways.

---

[2] Thanks to Paul Skelly, formerly of ASL, GTE Labs, who made the necessary modifications.

## SS7 NETWORK EMULATION AND CONNECTIVITY

As described earlier in this chapter, we used an Eagle (from Tekelec; see www.tekelec.com, 2001 for details) STP to provide SS7 functionality, and connectivity to the CLASS-5 switch (GTD-5 switch) and to the softswitch. Usually, one or more SS7 access links (A-links) [2] are used for both connections. The STP supports both V.35- and DS1/T1-type interfaces for the SS7 links. However, it is possible to support other interfaces, such as DS0A and fractional T1, by using appropriate interface converters. In addition, by using CSU/DSU and proper wiring, it is possible to offer connectivity to the PSTN switches, which are located hundreds of yards away.

## NETWORK TIME SERVER

The network time server (NTS), shown in Figure 5-1, is basically a specialized server that provides timing information to all IP domain network elements. The NTS can derive the clock from an SS7 network domain element (e.g., from the STP) or it can have its own clock source derived from a GPS receiver, for example. As mentioned earlier in this chapter, TrueTime's (www.truetime.net, www.truetime.com, 2001) NTS 100, 150, or 200 model server can be used in the testbed. This helps the critical IP domain network elements (such as IP-PSTN GWs, CC, and softswitch) and applications clients (such as IP phones) to synchronize precisely with the NTS over an emulated IP WAN. Without proper synchronization of the critical network elements, measurements of time delay (or latency) and delay jitter across the network would not be accurate; consequently, it would be difficult to monitor and maintain the desired level of service quality. Lack of synchronization may also result in inaccurate recoding of call start and stop times, thereby generating erroneous CDR files.

## TELEPHONE CALL EMULATION SUITES

We developed HVB-based telephone call emulation suites. Hammer (www.hammer.com, www.empirix.com, 2001) is a WindowsNT server-based PSTN call analysis and bulk call generation system [3] that can accommodate, up to six T-spans. It uses AG-T1 cards from Natural Microsystems (www.nmss.com) and can support CAS, ISDN, and SS7 protocols. We used the CAS and ISDN interfaces in our testbed. Test scripts and test suites that have been written using HVB can be used for call-progress analysis and bulk call generation. It is also possible to schedule repeated running of the same test suite at a predetermined frequency over a set of incoming and outgoing channels of the Hammer tester. In addition, recently Hammer has added ITU-T's P.861 standard-based voice quality measurement using the perceptual speech quality measurement (PSQM; 0: best match and 6.5: worst match) technique. Other

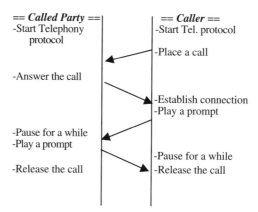

**Figure 5-2** Sequence of interactions during a typical telephone conversation.

techniques for objective speech quality measurement include the perceptual analysis/measurement system (PAMS), available in the digital speech level analyzer (DSLA) products from the Malden Electronics Ltd. (www.malden. co.uk, 2001). PAMS brings in the effects of perceptual relevance in speech signal recognition efforts.

The model of a test telephone call can be described as follows. In a telephony/conversation session, there are two or more interacting players: for example, a calling party, a called party, a local switch, and a voice response unit (VRU). In Hammer testing, a conversation is emulated by using a test suite that consists of at least two HVB scripts; one emulates the caller and the other emulates the called party, with communications occurring over the line or channel (over the Intranet) under test. Figure 5-2 is a simple ladder diagram showing the sequence of interactions between the two HVB scripts playing the roles of caller and call receiver. Note that the sequence of play prompt and pause can be executed a number of times in order to increase the length of the emulated call. This basic call setup method can be enhanced to perform a multistage call setup using personal identification number (PIN)-based caller authentication, as shown in Figure A-2. Similarly, it is possible to add a pre-specified amount of precall waiting time between each call, as shown in Figure B-3, and to stagger the calls or connection requests, as shown in Figure B-5.

For IP telephony tests and measurements, it appears that most of the Hammer's built-in call-progress time/tone detection functions cannot be directly utilized. Therefore, we have decided to use the tone detection procedure with the tone's duration and tolerance frequencies adjusted empirically, per the implementation in the IP-PSTN GWs. Furthermore, when making large numbers of simultaneous calls, sometimes it is necessary to add a precall wait time; otherwise, the call establishment attempts fail repeatedly. This happens because of limited processing (CPU) capacity in the implementation of IP-PSTN GWs.

Hammer also provides VoIP test suites that include connection testing, voice prompt testing, DTMF testing, and load testing suites. However, during our test phase, we found that their test suites are neither stable nor robust enough to handle the variety of single- and multistage calls that we needed to evaluate the emerging IP-PSTN GWs.

We have developed HVB-based test suites and an oscilloscope-based setup for measuring postdialing delays and one-way voice transport delay. In addition to measuring the DTMF digit transmission performance, we have developed a variety of test scripts and suites (presented in Appendixes A, B, and C) to determine the call setup performance of the GWs under test. All of the test suites use version 2.1.3 of Hammer's operating system (HammerIT) software.

It is also possible to use Hammer's VoIP test suites to measure one-way voice latency and to derive the voice quality score via PSQM—as defined in ITU-T's P.861 recommendation—measurement of voice transmission quality. For calibrating the results obtained by using Hammer's VoIP suites, we used the results obtained from the oscilloscope-based setup and measurements.

## EPILOGUE

The testbed presented in this chapter has been used to study the basic interoperability of IP and PSTN domain call controls and transmission of real-time voice traffic over an emulated IP network. It is possible to enhance the capabilities of this testbed. The enhancements depend mostly on the test objectives. For example, additional network elements can be incorporated to investigate a variety of other interoperability scenarios. These may include interoperability of (a) a variety of VoIP call control and signaling protocols, (b) calls between IPv4 and IPv6 [4] domains, (c) calls between ANSI and ITU-T SS7 domains with one or more IP domains for real-time voice transmission, (d) services and call features invoked from the IP domain by the PSTN clients and vice versa, and so on.

## REFERENCES

1. R. J. Bates and D. W. Gregory, Voice and Data Communications Handbook, McGraw-Hill Book Companies, New York, 1998.
2. T. Russell, Signaling System #7, Second Edition, McGraw-Hill Book Companies, New York, 1998.
3. S. Gladstone, Testing Computer Telephony Systems and Networks, Flatiron Publishing, Inc., (now CMP Books) New York, 1996.
4. C. Huitema, IPv6—The New Internet Protocol, Prentice Hall, Upper Saddle River, New Jersey, 1998.

# 6

# VoIP DEPLOYMENT IN ENTERPRISES[1]

Enterprises are probably the first ones to derive the benefits of running real-time telephony and associated services over an IP network. Before the advent of VoIP, enterprises generally had phone lines for real-time voice and fax services, and a data network based on dial-up, X.25, frame relay (FR), ATM, IP, and so on for data communications services [1]. IEEE standard 802.3 protocol or Ethernet-based LANs are very common in enterprises [2] for data communications networking.

Small, medium-sized, and large enterprises can be defined as follows:

- The small office home office (SOHO) usually supports a few (fewer than eight) phone lines and a small (fewer than 16 ports) LAN. Small enterprises commonly support a few (fewer than 16) phone lines and a small LAN (about 32 ports). They are usually confined to one to four geographical locations.
- Medium-sized enterprises usually need tens of phone lines, a router, and multiple (medium-sized LAN of 32 to 64 ports) Ethernet switch-based LANs per location. Typically, they consist of a few offices in multiple geographical locations.
- Large national enterprises usually need tens to hundreds of phone lines and multiple large Ethernet switch- and router-based LANs per location. Typically, they consist of tens of offices in multiple geographical locations.

---

[1] The ideas and viewpoints presented here belong solely to Bhumip Khasnabish, Massachusetts, USA.

The introduction of VoIP in enterprises not only leads to convergence of multiple disparate networks in one physical infrastructure running only one (i.e., IP) protocol, it also opens up the network for delivering several new and emerging productivity-enhancing IP-based applications and services to the employees and customers of the enterprises. These new services include IP-based fax and conferencing services, unified messaging, find-me/follow-me services, Web-based call/contact centers, e-commerce and customer-care services, support of virtual or remote or tele-workers, and so on.

Although the operational and infrastructure cost savings are the prime motivations for incorporating VoIP services in enterprises, there are other factors that contribute equally to the decision. Some of these are (a) use of a uniform (i.e., IP only) service and network management platform throughout the corporation, (b) flexibility in service creation and maintenance using a Web interface, for example, and (c) simplicity in adding, moving, and changing the management of desktops/terminals within the corporation. In addition, it is often said that in medium-sized and large corporations, the investment in VoIP pays for itself within months (see, e.g., the case studies in the website at www. von.com, 2001).

The corporate IP network or the Intranet must be properly engineered so that it meets or exceeds the packet transmission delay jitter, packet loss, and packet transmission delay limits suggested in Chapter 4 and Appendix C. This will ensure the required level of quality, reliability, and availability of the VoIP service anywhere within the enterprise.

This chapter briefly discusses the required network endpoints, interfaces, and network elements for deploying VoIP in enterprises. It also presents some networking scenarios that can help corporations to migrate from offering traditional circuit switch-based telephony (e.g., centrex, PBX) services to its employees to delivering IP- and VoIP-based advanced and integrated communications services to its customers (for e-commerce applications) and employees alike.

## IP-BASED ENDPOINTS: DESKTOP AND CONFERENCE PHONES

IP phones are POTS or ISDN phone-like devices based on PCs, intelligent digital signal processors (DSPs), and real-time operating and networking software/systems. These devices are used for accessing real-time voice communications services from, for example, any communications application service provider (CASP) and for transporting real-time voice signals over the IP-based packet communication networks. Although the first-generation IP phones supported only G.711-based voice coding and proprietary or H.323- or MGCP-based signaling and call control, the emerging IP phones are supporting G.729-, G.726-, and G.723-based coding options, and are predominantly using the SIP protocol for call control and signaling. Many IP phones have built-in multiport Ethernet hubs to support seamless connectivity to LANs, and

**TABLE 6-1   Typical Features and Functionalities of IP and SIP Phones**

| | |
|---|---|
| Direct dialing based on digit | Music on hold |
| Direct dialing based on e-mail address | Caller ID blocking |
| Digit map support | Call forward |
| Private network dialing plan support | Anonymous call blocking |
| Direct inward dialing (DID) | Multiple directories |
| Direct outward dialing | Integrated multiport Ethernet |
| Call forward network | DNS service |
| Do not disturb (DND) | Inline power (over category 5 LAN cable) |
| Conferencing (four or more parties) | 10-BaseT and 100-BaseT |
| Call transfer with consultation | Auto-identification (easy add/move/change) |
| Call transfer without consultation | G.711, G.729, and wideband CODECs |
| Call waiting | Intercom support |
| Speakerphone with mute option | Plug and talk feature |
| Infrared port | Register station by using proxy |
| Adjustable and custom ring tone | In-band DTMF transmission |
| Hearing aid–compatible handset | Out-of-band DTMF transmission |
| Volume control | Local or remote call progress tone |
| Independent volume control | Network startup via DCHP |
| Last number redial | Date and time support via NTP |
| Display contrast control | Third-party call control via delayed media play |
| | |
| Internal phone browser | Support for endpoints in SDP |
| Call log | Local directory, conference call log |
| Call log filter | Message waiting indication (MWI) |
| Customizable display screen | Speed dial to voice mail box |
| Online help | General-speed dial |
| External speaker jack | Capability to add new applications |
| JTAPI support | Click to dial from outlook |
| Call hold | Access to application portal |
| LDAP-based phone book | Support of QoS by packet marking |
| Presence management | Call park |
| Vcard exchange via phone | Barge-in calling |
| Video streaming | Intelligent attendant |
| Scanning/checking e-mail | Rolodex-style scroll knob |
| Display call image | Automatic version update (via TFTP or HTTP) |
| Embedded Java | Ability to view video graphic files |

are also capable of deriving electric power using the same Ethernet cable (a category 5 cable) that they use for connecting to the Ethernet LAN hub/switch. Table 6-1 presents a list of features and functionalities that are commonly available in IP and SIP phones. It appears that these phones are capable of supporting many of the productivity-enhancing features and functions that are commonly used in the business communications environment. Also, since IP phones facilitate dynamic registration of clients (or endpoints) via the dynamic host configuration protocol (DHCP) features of IP, they make adding, moving, and changing very simple. Finally, since IP phones use the same data networking infrastructure and technologies, they make enterprise network evolution and management more seamless and less expensive.

A number of recently developed IP phones support conferencing features and functions that are commonly available in expensive traditional PBX phones or in the phones that can only be purchased as part of the key telephone systems (KTSs). These IP phones offer full-duplex audio, display functions, and features such as access to voice mail and name directories, call add, drop, and transfer, and interconnecting multiples conferences bridges. In addition, these conference IP phones can be used as a client to the IP-PBX (described in the next section) in integrated voice (TDM) and data (mainly IP) networks by simply plugging them into the LAN, or Ethernet network [1,2] jack in any conference room in the office.

Many companies, including Cisco (www.cisco.com, 2001), Pingtel (www. pingtel.com, 2001), Polycom (www.polycom.com, 2001), and Siemens (www. siemens.com, 2001), have recently started marketing their desktop and conference IP phones to high-end residential and enterprise markets.


## IP-PBX, IP CENTREX, AND IP-BASED PBX TIE LINES

IP-PBXs are PBX devices that support the following:

  a. Various IP telephony and/or VoIP features;
  b. Call processing/control and attendant features/functions that are available from traditional circuit-switched PBXs;
  c. One or more of the following types of phones: analog, digital, ISDN-BRI, IP, and so on; and
  d. One or more T1/E1-CAS/PRI links and digital subscriber lines (DSLs) for connectivity to PSTN switches and IP trunks for local and/or wide area data/packet networking. The IP trunks can be used to interconnect the IP-PBXs of a corporation in different geographical locations over an IP-based corporate virtual private network (VPN).

Deployment of IP-PBX not only reduces the costs and enhances the features and capabilities of enterprise communications, it also simplifies the software upgrading and management of the integrated voice and data infrastructure. In addition, IP tie lines or IP trunks can be used to interconnect the IP-PBXs in different geographical locations. The use of IP tie lines can (a) make the same advanced call control features of the corporation's headquarters available to employees in remote branch locations and (b) allow employees to hold conference calls over a wide geographic area, avoiding long-distance telephone charges.

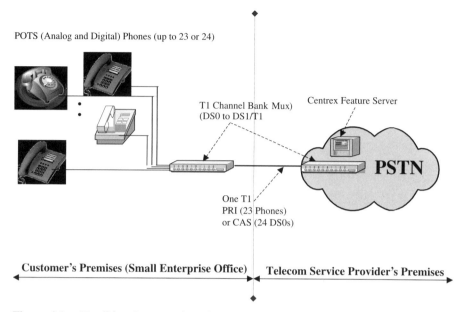

**Figure 6-1a** Traditional centrex-based telephone service offering to enterprise or corporate customers.

IP-PBXs can offer the same set of services that traditional analog centrex and ISDN centrex offer. In analog centrex and ISDN centrex, the call control features and functions reside in the CLASS-5 switch placed in the central office (CO) building, with, for example, a dedicated T1 line for every 23 (for T1-PRI) or 24 (for T1-CAS) telephone terminals on the customer's premises, as shown in Figure 6-1a. This system is not only expensive to maintain, it also may offer only a limited and/or proprietary set of centrex features. In PBX (traditional) or IP-PBX (emerging), these functions are usually hosted in the network elements that reside on the customer's premises, and one or more T1 (traditional) or DSL (emerging) connections to the CO can be used for PSTN connectivity, as shown in Figure 6-1b. The DSL connections can carry both voice and data traffic over the same link and are usually significantly less expensive to maintain than T1 connections. Also, since the call control can be local and IP-PBX supports Internet connectivity, it is not necessary to have one T1 line for every 23 (for T1-PRI) or 24 (for T1-CAS) telephone terminals on the customer's premise (discussed more in the context of Figure 6-3 at the end of this section).

Note that with the advent of VoIP and the ubiquitous availability of IP-based network connectivity, analog centrex and ISDN centrex are evolving toward IP-based centrex. To offer IP-based centrex services, the service provider needs to support a high-quality (i.e., with guaranteed QoS) broadband (over DSL, T1, Ethernet, etc.) IP link to the customer's site, instead of offering expensive T1 lines that support voice calls only. The customer can use the

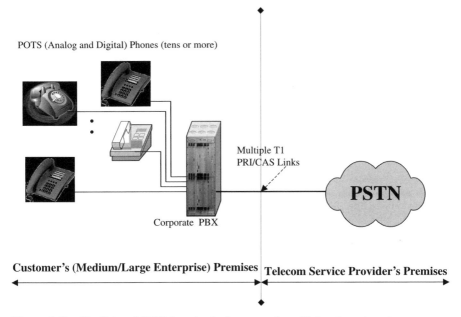

POTS (Analog and Digital) Phones (tens or more)

Multiple T1
PRI/CAS Links

**PSTN**

Corporate PBX

**Customer's (Medium/Large Enterprise) Premises** | **Telecom Service Provider's Premises**

**Figure 6-1b**    Traditional PBX-based telephone service offering to enterprise or corporate customers.

broadband IP link for simultaneous transmission of voice and data traffic to deliver a variety of enhanced applications and services to employees. To support legacy telephones and fax machines, customers need an IP-PSTN GW on the premises. This GW provides signaling and media (bearer traffic) conversion from the legacy TDM domain on the customer's premises to the IP domain in the service provider's CO. This conversion helps communications with appropriate network elements like the IP-PSTN GW, VoIP CC, softswitch, and so on in the IP network of the service provider. Note that IP PBX and IP centrex offer a superset of the traditional analog centrex and ISDN centrex services, some of which are shown in Table 6-2 (further details can be found at www.ip-centrex.org/features/index.html, 2001). Table 6-3 presents typical IP telephony and VoIP-related features expected from IP centrex and IP PBX. Additional autoattendant and CC-related features that are expected to be supported by IP-PBX-like devices are shown in Table 6-4 and discussed in the next section.

When IP-PBXs are used, enterprises can install the IP telephony network elements or devices adjacent to the data-networking (e.g., LAN) infrastructure, reducing wiring and management complexity and physical footprint requirements [3]. Also, IP-PBX supports not only the flexibility and efficiency of IP telephony, but also peer-to-peer VoIP connectivity over LANs and WANs. In addition, the IP domain network elements use open (or standards-based) and Web-based interfaces for call control and feature/service provisioning and management. Consequently, it is relatively faster and simpler to manage soft-

**TABLE 6-2    Typical Call Control Features and Functionalities of Traditional Centrex and PBX**

| | | |
|---|---|---|
| Automatic call-back (Camp on) | Intercom | Message- and/or music-on-hold |
| Bridged call appearance | Last number redial | Free seating |
| Call forwarding (internal and external) | Message waiting (using light and/or tone) indication | Time-of-day (e.g., night)–based service |
| Call pickup | Multiple call appearance | System speed dialing |
| Caller ID display and called ID blocking | Mute | Voice mail |
| Hunt groups | One-button speed dial | Call trace |
| Distinctive ringing | Call transfer | Call park |
| Call drop | Volume control | Call conferencing |
| Call hold and waiting | Automatic alternate routing | Do not disturb (DND) |
| Auto redial and auto call back | Automatic route selection (for outside or 6+, 7+, 8+, 9+, etc. calls) and auto-direct connect | Interactive voice response (IVR)–based service and recorded announcements |
| 700/900 call blocking | Call screening and blocking | Emergency call attendant |
| Call join, fork, stack, etc. | Automatic detection of fax tone | Call intercept treatment |

ware upgrading and to roll out new service features (e.g., unified messaging, find-me/follow-me services) across the enterprise.

Both traditional PBX vendors and Internet router manufacturers are developing and marketing IP-PBX and other relevant feature GWs and application servers. Some of them are Avaya (www.avaya.com, 2001; formerly a part of Lucent), Nortel (www.nortelnetworks.com, 2001), Siemens (www.siemens.com, 2001), NEC (www.nec.com, 2001), Mitel (www.mitel.com, 2001), and Cisco (www.cisco.com, 2001). Note that some of the commercially available IP-PBXs can support many new and emerging services in addition to tens of call processing features and functions that are available in traditional circuit-switch-based PBXs.

Figure 6-2 shows possible architectures for migration of traditional centrex services to IP-based centerx services with minimal infrastructure investment by customer but a somewhat significant (less for ISPs but perhaps more for telecoms) capital investment from the service provider. Details of the costs depend on interface and service requirements, scope of the deployment, age of the equipment (handsets) and the IP network infrastructure already in place, and so on, and can be evaluated on a case-by-case basis. IP centrex customers can add new endpoints (phones) without requiring new phone lines to the telecom service provider's central office, and also can roll out many new and advanced IP-based services in a customized fashion just by adding new servers to their local IP network (LAN or Intranet). Many existing telecom switch manufacturing vendors are developing either (a) line cards that integrate with exist-

**TABLE 6-3    Typical VoIP and Related Features and Functionalities Expected from IP-PBX and IP-Centrex**

| | | |
|---|---|---|
| Simultaneous support of IP and POTS (analog, digital, ISDN-BRI) phones | Support of VoIP for both access (IP phones) and transport (inter-PBX IP trunk) for toll bypass | Support of a large number (tens, hundreds, thousands) of IP phones |
| Support of self- and Web-based configuration, provisioning, user profile management, and so on for easy add/move/change, find-me/follow-me, and other services | Support of the line-card-based (or integrated) VoIP GW | Support of QoS in both access and transport domains by using access control and by marking the VoIP packets as high-priority packets |
| Support of the existing and emerging VoIP signaling and call control protocols (e.g., H.323, MGCP, SIP) | Support of a wide variety of voice compression schemes (e.g., G.711, G.729, G.723) with and/or without silence suppression | Support of electronic numbering (IETF's ENUM, RFC 2915/16) to enable dial using the e-mail address, URI, URL, and so on |
| Support of automatic fallback to PSTN trunks for call routing when the IP link(s) are congested | Support of unified messaging including real-time and store-and-forward fax transmission service | Support of security, scalability, reliability, and emergency call routing |
| Support of IP-VPN and voice-VPN services | Support of instant messaging, meet-me/follow-me conferencing (audio and video), and so on | Virtual enterprise, integration with e-mail (MS-Outlook, MS-Exchange, Lotus Notes, etc.), presence management, and so on |

ing devices to support the required interfaces and functions or (b) GW devices to support feature and service interaction and transport mediation between IP and PSTN domain networking and service delivery elements.

Figure 6-3 demonstrates how an existing circuit switch-based PBX infrastructure can be migrated to an IP-PBX-based one by adding an embedded VoIP CC and GW (to PSTN) line card in the existing PBX. Another option for such a migration would be to use a separate physical device that functions as an integrated VoIP GW and call controller or proxy of a separate CC, depending on the system architecture.

Although there are a number of protocols (H.323, SIP, MGCP, etc., as discussed in Chapter 3) for controlling IP-based endpoints (e.g., a phone), it appears that because of its openness and simplicity, IETF's SIP is enjoying

**TABLE 6-4   System Features and Functionalities Expected to Be Supported by an IP-PBX**

| | | |
|---|---|---|
| Automatic call distribution (ACD)–based call control (including priority queueing) | Display of call duration and distribution | Call and call transfer between seats (positions) |
| Attendant override or barge-in (including automatic station relocation) | Supervision and monitoring of calls | Direct inward and outward dialing |
| Call display and ANI/DNIS-based service | Recalling a call | Call detail recording (CDR) |
| Route and trunk group selection (automatic or manual) | Support of computer and telephony integration (CTI) | Emergency access and night service |
| Fax mail (single or group, internal or external, etc.) | Programmable toll restrictions | Voice mail and/or video mail–based call back |
| Priority and serial calling | Station hunting | |

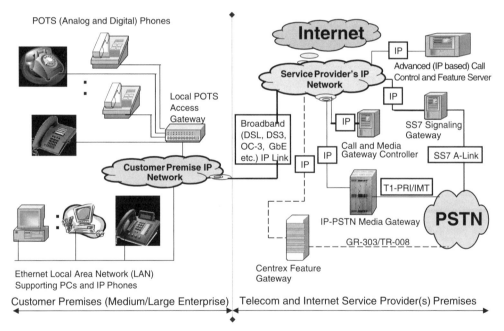

**Figure 6-2**   Evolution of a traditional centrex service offering to IP and technologies-based centrex service delivery. The connections shown by the dashed line are required when PSTN call and feature control reside in the PSTN network, and (a) SS7 SG, call and MGC, and (b) advanced feature server are not deployed. The centrex feature GW supports the GR-303/TR-008 interface to the PSTN and may contain the VoIP CC and MG.

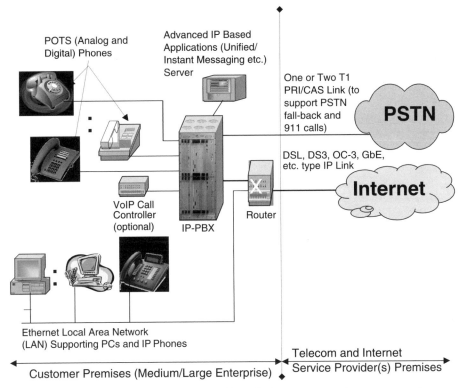

**Figure 6-3** IP-PBX-based telephone service offering to enterprise or corporate customers.

significantly more support from both standardization organizations and vendor communities. And for controlling the VoIP GW devices from the CC (or call manager or call server), the MGCP and Megaco/H.248 (discussed in Chapter 3) protocols are becoming clear winners.

## IP-VPN AND VoIP FOR TELE-WORKERS

VPNs use leased telecommunications links or shared Internet trunks to provide point-to-point private logical channels for data and/or voice communications. The flexibility and ubiquity of IP have motivated many Internet and telecom equipment manufacturers to develop IP-based virtual private networking (IP-VPN) devices that can support integrated real-time voice (using VoIP) and data services over broadband IP links. The broadband IP link—shown in Figure 6-4—could be a digital subscriber line (DSL), a cable modem-attached CATV line, a wireless or Ethernet local loop or IP over asynchronous transfer

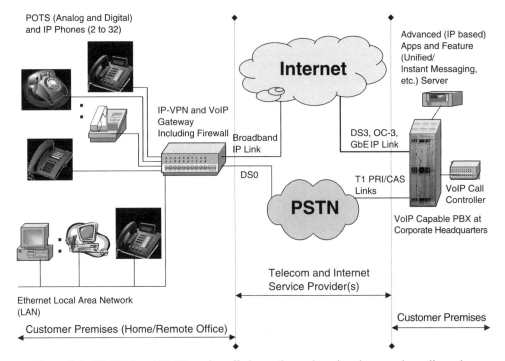

POTS (Analog and Digital)
and IP Phones (2 to 32)

Internet

Advanced (IP based)
Apps and Feature
(Unified/
Instant Messaging,
etc.) Server

IP-VPN and VoIP
Gateway
Including Firewall

Broadband
IP Link

DS0

DS3, OC-3,
GbE IP Link

T1 PRI/CAS
Links

VoIP Call
Controller

PSTN

VoIP Capable PBX at
Corporate Headquarters

Telecom and Internet
Service Provider(s)

Ethernet Local Area Network
(LAN)

Customer Premises (Home/Remote Office)

Customer Premises

**Figure 6-4**   IP-VPN and VoIP service offering to home-based and remote/traveling tele-
workers.

mode/synchronous optical network (ATM/SONET) over fiber or laser, and so
on [1,4,5]. Many large corporations are setting up IP-VPN (and VoIP) services
in the Intranets to enable their valued employees, call center agents, and after-
hour attendants to work from home (telecommute) when needed. The main
benefits of using VoIP over IP-VPN are that the tele-workers (at home or at
any other location) can use the same telephone sets, and can have access to the
same sets of call features and services that they enjoy while working physically
in their offices—be it headquarters or branch offices.

Commercially available SOHO IP-VPN devices and IP-PBXs also offer
graceful fallback of the VoIP calls from the IP network to the PSTN networks
(using DS0 or BRI lines) so that the employees or users can enjoy the same
QoS including network-level availability and reliability. Similarly, local and
emergency (e.g., 911) calls can be routed through the local analog or BRI line
to the PSTN network. This allows efficient, low-cost, appropriate delivery of
calls to the local endpoint (another PSTN phone) or the public service access
points (PSAPs) as required.

As with all IP-based services, the major issues in offering IP-VPN and
related services using a logically overlaid private network over the public

Internet are that (a) security of the services needs to be maintained and (b) the QoS needs to be guaranteed for the application in question [1,2]. To satisfy the security requirements, it is necessary to offer authentication, encryption/ decryption, tunneling, and stateful firewall services to the endpoints over IP. Similarly, to ensure the QoS requirements, it is necessary to offer access control and bandwidth management services to the IP packet streams as they pass through the IP network (Intranet or the public Internet). Therefore, any feasible IP-VPN device must support these security and QoS requirements by using the appropriate software and hardware embedded in it.

Currently, a number of IETF recommendations and ITU standards are available to implement security and QoS maintenance services. For example, (a) public key infrastructure (PKI)–based digital certificates, a combination of static, dynamic, and biometric information or password, and lightweight directory access protocol (LDAP)–based remote authentication dial-in user service (RADIUS) for authentication/authorization/accounting (AAA), and so on can be used for user or endpoint authentication; (b) the IETF's IPSec with a multiple-digital encryption standard (DES; triple DES is very common) based encryption of messages with a large (e.g., 128 bit) key can be used to maintain privacy and secrecy; and (c) a header compression- and encryption-based point-to-point tunneling protocol (PPTP), layer-2 tunneling (L2TP), and so on can be used for information tunneling service; and (d) a TCP/UDP port, IP address, type of protocol, service, interface, and so on based packet filtering, stateful packet inspection, auditing, service logging, network address translation (NAT), and others can be used for firewall services. In order to maintain the desired QoS requirements, IETF's differentiated services (DiffServ), integrated services (IntServ), random early discard (RED), resource reservation protocol (RSVP), multiprotocol label switching (MPLS), and other protocols can be used. These are discussed in Chapters 2 and 3.

## WEB-BASED CALL AND CONTACT CENTERS

Web-based call and contact centers not only support all the economic and operational advantages of VoIP, they also offer the flexibility and other benefits of IP telephony. Traditional circuit-switched PBX and automatic call distribution (ACD)–based call and contact centers can be upgraded to support VoIP and Web-based management and control by adding a VoIP GW and an IP interface along with the required software. Many traditional PBX and ACD manufacturers—such as Avaya (www.avaya.com, 2001; formerly a part of Lucent), Nortel (www.nortelnetworks.com, 2001), and NEC (www.nec.com, 2001)—have already started working in this direction, and recently have started marketing their Web-based call and contact center products.

Support of IP telephony and VoIP in the call center makes adding, moving, and changing of stations and invoking of remote (offshore or home-based) call agents simple and affordable. In addition, by using ANI/DNIS and instant retrieval (over IP network) of up-to-date customer information, the call agent's interaction with the customer can be made as personalized and current as possible at the lowest possible cost. In the case of multisite call centers operating in multiple time zones, the interworking of the VoIP GWs in different call centers using intersite IP links makes centralized messaging and management of services inexpensive and efficient. In addition, since IP telephony supports open telephony and intelligent networking (IN) application programming interfaces (APIs) like Java API for intelligent networking (JAIN), Paraly, and TAPI/JTAPI, many of the required sales automation and inventory management (for e-commerce applications), trouble ticketing, and accounting software packages and servers can be developed and integrated easily and cost-effectively with the main customer care and customer resource management (CRM) system. To guarantee the privacy and security of electronic transactions (for e-commerce applications), the authentication, encryption, and firewall mechanisms discussed in the previous section (IP-VPN and VoIP for tele-workers) can be utilized.

The support of Web-based call control and ACD management also facilitates seamless availability of intelligent control, routing, and smart management of calls from any location within the enterprise, either over the corporate Intranet or remotely over the IP-VPN by using VoIP. These attributes also enable real unification of all types—voice, data/e-mail, graphics/fax, and so on—of messaging over one application for delivery over one IP-based network to one multimedia PC or an IP phone (with a built-in large display). Consequently, it becomes relatively simpler and easier for corporations to develop and integrate/launch many of the emerging IP- and VoIP-based advanced services. These services include Web-based internal (among the call center agents) and external (between an agent and a customer) collaborations, open browser-based network and service management, chat and instant messaging, Web-based clicking for making a call and scheduling a call back to a customer, and so on.

Web-based contact centers can be used to implement a virtual call or contact center by exploiting the same IP client-server-based architectures [1,2] that are commonly used for advanced IP telephony call control and service delivery. These types of contact centers can not only span multiple times zones over different geographical areas all over the world, they can also support nonstop (i.e., 24 hours/day, 365 days/year) customer services cost-effectively. For example, a customer calling for help or service at 9 P.M. Eastern Standard Time to a Boston-based call center can be routed over the corporate IP-VPN (supporting VoIP) to an offshore (e.g., based in India) call center where it may be 7 A.M.. It is also possible to train different overseas groups of call agents as specialists in different types of services, and incoming calls can be routed based on customers' category and requirements—as identified by ANI/DNIS and other information collected via the interactive voice response (IVR) system.

## NEXT-GENERATION ENTERPRISE NETWORKS[2]

Intranets are corporate or enterprise networks that facilitate seamless communications and networked computing within a single corporation. To carry out business functions, however, corporations have traditionally employed several other networks to provide business services such as telephony, faxing, computing, and network administration. Today, these services revolve around two basic networks: PSTN, which provides basic telephone services, and the Internet. As shown in Figure 6-5a, these separate networks handle telephone and data/Internet services today. Next-generation enterprise networks (NGENs) must fulfill current expectations for corporate networks, and must also seamlessly support mobility and multimedia applications. And, they must achieve this by using flexible, often adaptive, self-configuring, and interoperable architectures. NGENs also must be user-friendly and scalable. They must support bandwidth and QoS requirements, in addition to delivering the expected reliability and availability. To meet these requirements, the trend is to consolidate these disparate networks (PSTN and the Internet) into one simple (IP only) high-capacity, reliable, scalable, QoS-aware network, as shown in Figure 6-5b.

NGENs will use one set of protocols—such as IPv4 (which needs IPSec and QoS mechanisms to support the security and service requirements) or IPv6 (which has a larger address space, as well as built-in QoS and security support)—processes, and vendors and one set of management, administration, and billing processes. They must be able to easily accommodate emerging computing and communications technologies and next-generation unified and simplified services.

### Customers' Expectations

Corporations expect NGENs to seamlessly support automated services and applications, facilitate process reengineering and technology consolidation, and support efficient network management and maintenance. A few of the emerging strategies for consolidating PSTN and Internet-based networks are as follows:

- *IP telephony* in the form of ID-aware voice plus data terminals (e.g., with a built-in multiport Ethernet hub), which includes supporting mobility in addition to facilitating add, move, and change activities.
- *IP-PBX or packet voice exchanges (PVXs) and integrated access devices (IADs)* to replace PBXs and terminals. IADs usually support DSL-based services and hence eliminate the need for multiple physical lines (DS0s)

---

[2] This section is based on the article "Next-Generation Corporate Networks," by B. Khasnabish, published in the IEEE IT Professional Magazine, Vol. 2, No. 1, pp. 56–60, January–February 2000.

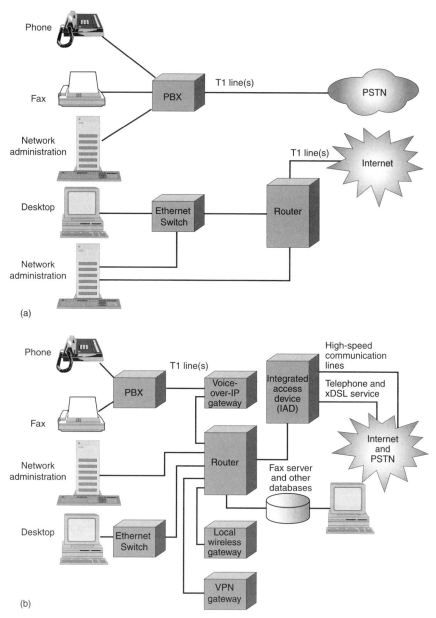

**Figure 6-5** Enterprise networks commonly have PSTN and Internet components, as shown in (a), but the next-generation enterprise networks will combine the two using the IP as the glue, as shown in (b).

from a central office to a customer's premises. One physical line from an IAD can support multiple virtual circuits. PVXs can handle both data- and packet-based voice services. Use of PVXs can avoid scalability and inter-operability constraints as well in certain scenarios.

- *Storage area networks (SANs)* are network-, server-, or micro/pico-area networks attached to a massive data-storage facility. They support the creation and control of services. SANs and mainframes are useful for data mining/warehousing, e-commerce, transaction execution, and peering. *Peering* allows networks to exchange traffic directly rather than using the Internet backbone; it permits a more efficient, seamless exchange of data. An effective SAN should support the lightweight directory access protocol (LDAP), billing, and semidistributed Web-based control and management. Software packages based on extensible markup language (XML) can facilitate interaction and information exchange over the Web without requiring massive rewrites or modifications of the existing/legacy systems/applications of the individual companies. Whatever the strategy is, it is important to consider support for the several functions that the integrated network must provide.

### Process Reengineering and Consolidation

Next-generation networks must support process reengineering and consolidation applications. Such applications include those for enterprise resource planning (ERP), e-commerce, data mining and data warehousing, and peering of servers and networks. Although the objective of implementing these applications is to reduce cost and complexity, nonjudicious use of these technologies can produce adverse effects.

### Proactive Maintenance

Performing proactive maintenance is becoming another function for corporations. This capability was "nice to have" in the 1990s, but it is rapidly becoming a necessity as networks become more complex. Examples of proactive maintenance are as follows:

- Software or system configuration and version-management applications using Desktop Management Task Force's Desktop Management Interface (DMI) or Sun's Jini;
- Remote/self-configuring and maintenance of desktop computers and applications; and
- Network and traffic configurations management using time- or traffic-pattern-triggered network and traffic management policies.

**Support for QoS**

Maintaining QoS calls for optimizing network access and traffic routing. A QoS-aware network recognizes various categories of data and attempts to guarantee an associated level of service, which is defined by parameters such as allowable delay, delay jitter, packet loss, and so on.

Emerging protocols include customer premises and/or desktop devices with native support of TCP/IP (transmission control protocol/Internet protocol) and UDP/IP-based (user datagram protocol/Internet protocol-based) access. UDP is commonly used to support real-time or delay-sensitive/loss-tolerant services over IP. The goal is to permit such devices to support varying qualities of service [6,7].

For IP-based services, QoS issues are currently being addressed by IETF's DiffServ, IntServ, and multiprotocol label-switching (MPLS) working groups, Internet 2's QoS Forum (www.internet2.edu, 2000), and other similar organizations such as QoS Forum (www.qosforum.com, 2000).

Essentially, all these organizations are attempting to develop parameters for the required *service-level agreement* (SLA). These parameters spell out the reliability, availability, response time, delay variations, and security necessary to satisfy an application's QoS requirements, such as real-time voice transmission, VPN, and so on. Enforcing SLA agreements requires tools for monitoring, configuration, and provisioning management.

Small to medium-sized organizations may want to outsource the operation of their corporate networks or may choose to co-source them by using both internal and external resources. This would help these businesses be more proactive in monitoring and managing traffic. Outsourcing can also help better predict service and capacity requirements for infrastructure planning and migration.

**Support for Multimedia**

Multimedia applications include the use of voice, video, and image clips in e-mail. To further complicate the situation, users are asking for unified messaging—anywhere, anytime access to voice mail, e-mail, and faxes over IP.

Groupware and other applications that support remote collaboration form another class of multimedia applications. Corporations use such applications for problem solving, service provisioning and management, videoconferencing, and Web- or CD-based employee training and distance learning.

The greatest technical problem in supporting multimedia services over IP is that real-time traffic (data or packets) must reach its destination within a preset time interval (delay) and with some tolerance of the delay variation (jitter). This is difficult because the original UDP/IP operates on a best-effort basis and permits dropping of packets on the way to a destination.

Critical non-real-time traffic—such as topology and routing-table update information—is loss-sensitive. The entire network could collapse if it loses *any* packets. The effective solutions [6] call for using IP with

- Preventive and/or proactive traffic management schemes at access, network, and nodal operation levels and
- Reactive traffic management schemes at nodal, access, and network operation levels.

Preventive control mechanisms at the access level use traffic descriptors, traffic contract, and conformance testing to exercise control. At the network level, sharing and/or spreading traffic across various routes to a destination is most useful for non-real-time traffic. At the nodal (queuing) operations level, judicious use of traffic shaping at the intermediate nodes can help network administrators perform traffic management and control.

Reactive control mechanisms at nodal operations call for discarding packets if the queue is growing quickly and the incoming packets are neither important nor urgent. At the access level, one can mark (using the IP type-of-service byte) or discard packets on the basis of port or connection type if oversubscription persists. At the network level, one must control the traffic flow rate in physical and virtual connections by using the route congestion information flowing back and forth between various source-destination pairs or patterns. To be effective, the reactive scheme requires a faster response/reaction time than the rate at which congestion is occurring.

For nonurgent, loss-sensitive traffic, the nodal buffer can be designed to be as large as needed without degrading traffic transmission performance or adversely affecting performance expectations at the application level.

For urgent or delay-sensitive traffic, a network must use suitable scheduling and/or cut-through routing. For example, for a given traffic profile and service discipline, one can calculate the buffer size for tolerating the loss of 1 in 1 million packets. However, for delay-sensitive traffic (such as voice or real-time video), both the number of intermediate hops and the nodal buffer space must remain small so that the packet transport delay and delay variation stay within certain limits. Adding more storage space (buffer) at the nodes would not solve the problem.

In order to minimize the maximum queuing delay, the network design should consider minimizing the number of *active* nodes crossed from source to destination. Consequently, the concept of virtual (private) networking comes into picture. In VPNs, paths are almost always preset, and the route characteristics are well guaranteed via SLA parameters negotiated with the service providers.

**Improving Wired Access**

Several access options are emerging for hardwired connections. Varieties of digital subscriber lines (xDSL) are popular because the infrastructure to support them is less expensive than that of other options. Corporations can also use cable-modem-based connections to link the LAN GWs in different geographical locations. Optical fiber connections are supra-high-speed options to

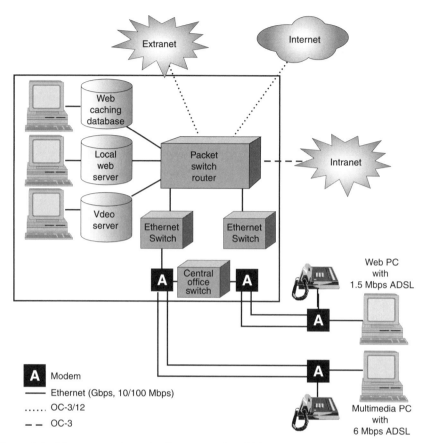

**Figure 6-6**   Next-generation wire-line networks within a corporation could use DSL to support Internet, Intranet, and Extranet services.

link corporate LAN GWs, but they are more expensive. A number of currently available and emerging high-speed access technologies are discussed in detail in [4,5]. One such service, which uses the asymmetric digital subscriber line (ADSL) technologies, is shown in Figure 6-6. This architecture can support local Web servers, Web caching, merchant services and e-commerce, the point-to-point tunneling protocol (PPTP) for secure LAN access, and multimedia services such as IP-based audio- and videoconferencing.

**Wireless Access**

For built-in wireless telephony/PBX, the base station should support the required mobility (handoff capabilities), QoS, and channel capacity. Standards for wireless telephony have arisen: the European CT-2 standard (CT, cordless telephony) supports approximately 8 handsets per base station, and the Digital

European Cordless Telephone (DECT) standard supports up to 12 handsets per base station over a 100- to 200-m diameter. A wireless PBX must also support users both within and outside a location [8,9]. The wireless connection must be secure, and the system must authenticate the user before allocating a channel or circuit.

To successfully integrate wireless services into a network, one should consider

- Capacity planning to support efficient operations today;
- Capacity and infrastructure planning for changing the network to support new applications and operations;
- Support of acceptable (less than 10%) blocking and handoff from the handset to the base station, which includes reduced blocking between the base station and wireless PBX, as well as acceptable blocking at the wireless PBX to the Intranet or PSTN (access-level blocking results in redialing for service, and in-transit blocking causes handoff failure); and
- Backups for unexpected events such as a flood or fire, which may cause facility outages.

Although the initial deployment of a wireless PBX can be expensive, it pays off in improved employee productivity and enhanced customer (internal and external) satisfaction.

For wireless communications within a single corporate network, it is prudent to consider an IP-based virtual network to support ubiquitous and uniform terminals and services. In this environment, network segments must interoperate seamlessly with public and private carriers. Additional operations, administration, and maintenance costs for managing such a worldwide virtual network also need careful analysis.

Other prime issues include discovering the called mobile unit or terminal for completing a connection request and maintaining the integrity of the offered connections or calls in progress. Systems typically accomplish this by paging, broadcasting, and/or mobility tracking or management methods. Tracking methods include

- Location and mobility tracking databases (home and visitor location registers), such as the ones used in the public PCS networks;
- Global positioning system (GPS) coordinates for tracking the location of a terminal and then using low-overhead mobility management techniques for maintaining connection continuity; and
- Various satellite-based systems.

An advantage of an IP-based global virtual network is that users can use the same handset either within the company's building or while traveling inside and outside the country.

**Enterprise Network Management**

Three issues need careful considerations in implementing emerging enterprise network management (ENM) options:

- *Interoperability*: Vendors and standards organizations are proposing a variety of architectures, platforms, and protocols. The more the types of networks the corporation has, the more complicated interoperability becomes.
- *Architectures*: Architecture needs to be assessed to determine its scalability and how well it deals with heterogeneous systems.
- *Synchronization*: Technological advances must synchronize with business needs.

For example, consider the SLA parameters for the PSTN. Required reliability for the PSTN is 99.999%—the "five 9s." A user must also receive a dial tone within 300 msec of picking up the handset 95% of the time. The PSTN must support an average holding time (call length) of 3 min or 180 sec. The PSTN's connection drop rate during a call is almost *zero* because it always has redundant paths and uses the latest released (and healthy or continuity-tested) circuit to set up a new connection. Next-generation enterprise networks need to provide PSTN-like services using IP for both signaling and media traffic transmission while still meeting the above reliability and availability requirements.

Emerging ENM strategies include the following:

- Management of VPNs, which use privately managed logical channels (or a mesh of channels) over public physical links for ENM;
- Virtual network management, which is a form of customer network management; and
- Management of the virtual enterprise—that is, the entire business and all networking processes are virtually defined. In this case, multiple levels of overlay can create complexity.

NGENs have two difficult responsibilities to fulfill. First, they must continue to provide the reliability and functionality now provided by older tested technology. Second, they must support future technology and service evolution. Doing both well could provide a make-or-break competitive advantage to a corporation.

## EPILOGUE

We have discussed the deployment of VoIP in enterprises—from desktop to centrex to PBX to call and contact centers and beyond. Following are some of the reasons why one should consider rolling out VoIP in the enterprise network:

- Converging the voice and data networking infrastructures;
- Bringing the integrated or converged network under the same set of management and maintenance portfolios (hardware, software, process, support personnel, etc.);
- Achieving savings on long-distance phone bills, because IP-based intersite (inter-IP-PBX) trunks can be used for calls between sites;
- Supporting unified messaging and making the call control and management features available uniformly across the corporation, irrespective of the location—headquarters, branch office, or remote/home office—where these services are hosted and from which they are accessed;
- Support of IP-VPN for remote workers and tele-workers, and of IP-based call, contact, and e-commerce centers for cost-effective, nonstop operations;
- Support of open APIs for cost-effective development/customization and integration of advanced IP- and VoIP-based call features and services; and
- Evolution or migration to an all-IP-based computing and communication infrastructure when resources such as black phones, leases on circuit lines (DS0s, T1s, etc.), PBX, and so on are fully depreciated.

However, there are also many issues that need to be carefully resolved before corporatewide availability of VoIP can be a reality. Some of these are mentioned below.

Circuit-switch-based network elements, links, and endpoints (black phones) are well known for their reliability, availability, and the QoS they provide. It may not be cost-effective to attempt to replicate the reliability, availability, and QoS values such as 99.999% of reliability/availability and a MOS score of greater than 4.0 for voice quality using shared-resource-based service protection. This is because the old paradigm of switched-resource-based service protection, underutilization (including the rule that 80% of traffic must remain inside and only 20% of traffic travel outside) based link capacity calculation and network topology design, and so on may no longer hold true when the same network carries voice, e-mail, fax, and messaging traffic in a unified fashion. Fortunately, there are a few remedies:

a. Design the corporate Intranet to always provide higher emission priority—for example, by marking the type of service (TOS) byte in the IP header—to real-time traffic such as voice packets from the traffic source on, as suggested in Chapter 2.

b. Overprovision the link capacity and/or enforce access control in both the LAN and the intersite IP links so that the real-time voice packets rarely suffer from nodal and links congestion.

c. Use independent IP links and PSTN links as contingency options (as shown in Figs. 6-3 and 6-4) for call routing during severe congestion due to faults or other unforeseen situations.

d. Consider deploying IP version 6 (IPv6, IETF's RFC 2460/1883, www.
   ipv6.org, www.internet2.org, 2001) based addressing, and other security
   and QoS offerings throughout the corporate network, or use IPSec along
   with the already deployed IP version 4 addressing-based network.

Many medium-sized and large enterprises are using these techniques to support VoIP services over their corporate IP network (Intranet).

Next, with the openness and flexibility of IP, the issue of maintaining
the security of services also becomes significantly more important. In circuit-
switched networks, switched or dedicated resources are used for communication, whereas in IP-based networks, computing and communication resources
are always shared to achieve optimum utilization of the network. This may
pose security threats from both inside and outside users/hackers. However,
there are many preventive mechanisms and good practices—as mentioned
earlier in this chapter—to minimize the risks of such attacks.

There are also regulatory issues that must be addressed to the extent to
which they are applicable within an enterprise. For example, the basic telephone service should be available even when the general supply of electricity is
not available. In centrex service, the PSTN service provider guarantees uninterrupted availability of service. When PBX- and IP-PBX-based services are
deployed, corporations themselves need to provide a backup power supply to
the PBXs and endpoints (soft phones and hardware-based IP phones) using
battery plants and/or an in-house electricity generator. Another issue involves
routing of emergency (i.e., 911) calls to the appropriate public service access
points (PSAPs), along with sufficient information to identify the location of the
endpoint (soft phone or hardware-based IP phone from which the call is being
made). When a soft phone or hardware-based IP phone is used for making a
911 call, a combination of information related to the (a) media access control
(MAC) address and/or IP address of the client phone, (b) LAN wiring diagram
and segment/subnet (which is serving the client phone) table, and so on can be
used to discover the location of the endpoint. These types of information are
usually available to the corporate IT department for network maintenance and
upgrade purposes, and can be made available on-line along with the DNS table
or entries in the DNS server or in the local network management console/
server. Hopefully, some best practices or standards will evolve within next few
years to resolve these issues.

The vision of an all-IP-based unified voice and data networking and computing system within an enterprise will become a reality only when real-time
voice/phone calls can be made over an IP network from one IP endpoint to
another using the e-mail or IP address, URL- or URI-based dialing, and so on,
instead of calling using the telephone number (ITU-T's E.164 address). IETF
has addressed this issue in a few RFCs (2806, 2915, 2916, 3026, etc.) to map the
telephone number into a naming authority pointer (NAPTR, RFC 2915) record using a DNS-based system architecture and protocol. The NAPTR record
contains a set of electronic numbers (ENUM) such as the e-mail address, wire

line and wireless phone numbers, fax number, URI, URL, and so on of the called party so that the calling endpoint can select the most appropriate identifier (ID) of the called endpoint. For example, if the calling party is a SIP phone, it may choose the e-mail address of the called party from the NAPTR record to make the connection request, that is, to issue an INVITE message using the called party's e-mail address. This message may travel over one or more routing information databases/servers—maintained by using IETF's telephony routing over IP or TRIP (RFC 2871) protocol—to ultimately deliver the message to the server hosting the destination IP phone.

A detailed discussion on the development of ENUM, its features, characteristics, interworking with IP telephony, call routing, PSTN's line information database (LIDB), and so on is available at ENUM (www.enum.org, 2001), the next-generation Internet (www.ngi.org/enum, 2001), and ITU-T (www.itu.int/osg/spu/enum, 2001) websites.

Depending on the VoIP- and IP-based services rollout strategy within an enterprise, these issues can be resolved in multiple phases over a sequence that best matches the budget, time, and infrastructure evolution plan. For example, small enterprises can migrate to DSL lines instead of paying for multiple DS0 lines to their premises. This will not only reduce their monthly phone bills, it will also allow them to harvest the benefits of using IP phones and many other advanced, productivity-enhancing call control and messaging features that are available for free or at nominal prices. Medium-sized and large enterprises can migrate from traditional centrex or PBX-based services to IP centrex and IP-PBX. This enables their employees to enjoy the flexibility and widespread availability of IP-based services from any location within the logical boundaries of their companies. Note that for medium-sized and large enterprises, it is important to introduce new networking and service delivery elements with a sufficiently granular target or long-term view of the network architecture. These new networking and service delivery elements must also support open or standard protocols and interfaces, such as SIP for IP telephony and messaging, IP for networking, Ethernet/gigabit-Ethernet/ATM for link layer communications, and so on. This strategy will not only help quick and cost-effective rollout of new and advanced services as dictated by the demands, it will also favor deployment of network and service delivery elements in the most convenient locations. The Multiservice Switching Forum (MSF) proposes one such architecture in their Release 1 implementation agreement, which is as shown in Figure 1-9.

## REFERENCES

1. W. Stallings, Business Data Communications, Fourth Edition, Prentice-Hall, Upper Saddle River, New Jersey, 2001.
2. V. Theoharakis and D. N. Serpanos, Editors, Enterprise Networking: Multilayer Switching and Applications, Idea Group Publishing, Hershey, PA, USA 2002.

3. B. Khasnabish, "Interior Design: Inside the Server Room," Network Magazine, Vol. 12, No. 11, pp. 105–109, November 1997.

4. B. Khasnabish, "Broadband To The Home (BTTH): Architectures, Access Methods and the Appetite for It," IEEE Network, Vol. 11, No. 1, pp. 58–69, January/February 1997.

5. D. Cuffie, K. Biesecker, C. Kain, G. Charleston, and J. Ma, "Emerging High-Speed Access Technologies," IEEE IT Pro Magazine, Vol. 1, No. 2, pp. 20–28, March/April 1999.

6. B. Khasnabish and R. Saracco, Guest Editors, Intranet Services and Communications Management, Special Topics Feature Issue of IEEE Communications Magazine, Vol. 35, No. 10, October 1997.

7. B. Khasnabish and M. Ahmadi, Guest Editors, Enterprise Network and Systems Management, Special Issue of the Journal of Network and Systems Management, Vol. 7, No. 1, March 1999.

8. Y.-B. Lin, B. Khasnabish, and I. Chlamtac, "The Wireless Segment of Enterprise Networking," IEEE Network, Vol. 12, No. 4, pp. 50–55, July/August 1998.

9. B. Khasnabish and M. Ahmadi, "Integrated Mobility and QoS Control in Cellular Wireless ATM Networks," Journal of Network and Systems Management, Vol. 6, No. 1, pp. 71–89, March 1998.

# 7

# VoIP IN THE PUBLIC NETWORKS[1]

VoIP technology is currently mature enough to be implemented in public networks (PSTN, cable TV [CATV], etc.), at least for long-distance telecommunications services to both residential and corporate customers. Either a private IP-based network (an Intranet) or an IP-based VPN can be used to guarantee the required QoS (call acceptance/drop rate, voice quality, etc.). In order to launch VoIP in the access loop, IP-based local access over digital subscriber line (DSL) or Ethernet in the first mile (EFM, IEEE P802.3ah) access, CATV networks, and wireless local loop (WLL) can be utilized. For corporate customers, the PSTN network can provide a variety of DSL-based access links to offer centrex features and functions and intersite IP-PBX connectivity, as discussed in Chapter 6.

In this chapter, we discuss evolution of various public network infrastructures (e.g., PSTN, CATV, etc.) to offer VoIP-based basic and advanced telephony services, either by using new IP-based network elements that are capable of supporting PSTN interfaces or by upgrading or modernizing the existing Telco-grade (i.e., the network equipment building system [NEBS]–compliant) PSTN elements with IP-based line cards, servers, and so on.

## IP-BASED TANDEM OR CLASS-4 OR LONG-DISTANCE SERVICES

In traditional PSTN terminology, if the calling and called parties are not served by the same CLASS-5 central office (CO) switch or cloud, then one or more

[1] The ideas and viewpoints presented here belong solely to Bhumip Khasnabish, Massachusetts, USA.

CLASS-4 or tandem-level switches and a transport network (see, e.g., Fig. 1-1 of Chapter 1) are required for establishing the connection between the two parties. That transport switch–based intermediate network constitutes a multi-connected and highly protected network that is commonly known as a *long-distance* (LD) or inter-LATA network, and the call becomes an LD call. In PSTN (circuit-switched) networks, to deliver high-quality voice, it is very common to use two-connected synchronous optical network (SONET) [1] ring-based transport networks with 50 msec of restoration time. PSTN networks use TDM-based circuit switching with a multiplexing hierarchy of DS0 (64 Kbps) to DS1 (or T1 or 1.544 Mbps), DS1 to DS3 (or T3 or 44.736 Mbps), and then OC-1 (51.84 Mbps) to OC-3 (STS-3 or STM-1), OC-3 to OC-12 (STS-12 or STM-4), and so on. Note that the DS0 to DS1/T1 multiplexer uses the byte interleaving technique, whereas the DS1 to DS3/T3 multiplexer uses the bit interleaving technique for multiplexing the information from the channels [1]. The requirement of 50 msec restoration time for transport was derived from the fact that any loss of information or fault with a duration of less than 50 msec in the transport network would not trigger any action—such as call drop or rerouting of trunks—at the lowest (T1 to T3 at the digital cross-connect system, etc.) multiplexing level. This also helps maintain the one-way end-to-end (ETE) delay of 150 msec, which is required to guarantee toll-quality (i.e., a MOS value of 4.0) voice signal transmission. This type of overprotection and overdesign guarantees both stability and higher-quality LD voice traffic transmission, but the cost of service is also very high (e.g., 25 to 30 cents per minute for a telephone call from Boston, Massachusetts, to San Francisco, California).

With the advent of VoIP, various next-generation LD service providers are deploying an IP-based transport network or leasing IP-based transport capacity along with the required network elements. These network elements interact with the transmission, call control, and feature servers of the PSTN network to deliver LD voice service—of varying quality—at a fraction of the cost of a telephone call from Boston to San Francisco. In addition, using appropriate shared redundancy, it is possible to achieve sub-50-msec restoration of transport services.

The customer can use 10-10-xxx based dialing, or they can dial a local phone number or a toll-free number (e.g., 1-800 or 1-888) to reach the desired IP-based call server. After proper authentication and authorization, the caller can proceed to dial the desired phone number for an LD call.

Table 7-1 presents a list of traditional CLASS-4 or LD services, features, and capabilities that the next-generation LD service providers need to support using an IP-based network, GWs, and service elements or servers. A detailed list of all of the CLASS-4 features and services can be found in the corresponding generic requirements (GRs) developed by Telcordia (www.saic.com/about/companies/telcordia.html, formerly Bellcore) for PSTN networks.

**TABLE 7-1    Traditional CLASS-4 and LD Service and Features**

Advanced intelligent network triggers (AIN 0.1 and 0.2 triggers)
Basic toll-free services like 1-800 and 1-888 dialing, national and international calling
    services, and so on
Caller ID and automatic identification of calling party's number (ANI)
Call/customer detail billing reports
Calling card service (prepaid and postpaid, with real-time update of balance)
Cellular Feature Group C and D trunk access ($+/-$)
Dialed number identification service (DNIS)
Emergency alternate routing within a prespecified time interval
Enhanced toll-free routing (e.g., NPA-NXX, time of day, day of week)
Feature Group B, C, and D SS7 trunk access
Feature Group B, C, and D multifrequency (MF) trunk access
Handling of ISDN user part (ISUP) and transaction capabilities applications part
    (TCAP) messages
Interface with SS7 network using A-F links
Interface with the interactive voice response (IVR) system
ISDN primary rate interface (PRI) trunk access
Local number portability (LNP) service
Routing of overflow calls, dial-around service using a four- to six-digit LD carrier
    selection code
Support of calling card fraud detection
VPN and software-defined network for voice VPN service
Wiretap service (communications assistance for law enforcement act [CALEA])
Zero $+/-$, 1+, etc. dialing for LD operator assistance and LD network access

## Elements Required to Offer VoIP-Based LD Service

Figure 7-1 shows one possible implementation of VoIP-based LD service that can be used as a model for gradual deployment of most CLASS-4 services. The required network elements are as follows:

a. IP-PSTN media gateways (MGWs) that interact with the PSTN network via access (e.g., T1-PRI/CAS) and trunking (e.g., intermachine trunk [IMT] with the speed of T1 or T3) links of CLASS-5-type central office switches;

b. An SS7 [3] SG that interprets the call setup and control messages from the SS7 network to the VoIP network, and vice versa;

c. A VoIP call server that controls the calls and IP-PSTN MGWs, and interacts with the billing system to capture the call detail records (CDRs) and put them in the appropriate format to generate customers' bills for the service;

d. Firewalls and other security enforcement devices (servers) to ensure that the calls originate from and terminate to the authorized endpoints, and

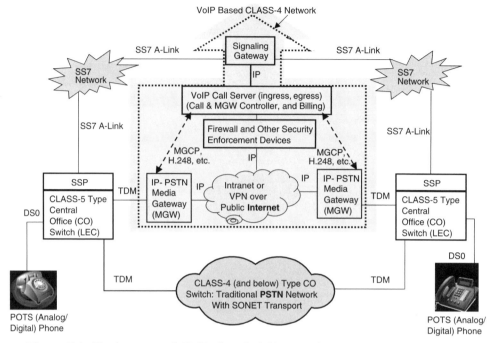

**Figure 7-1** Deployment of VoIP for CLASS-4 services (TDM: circuit-switched link, e.g., T1-CAS/PRI, T1/T3-IMT; IP: IP-based link; DS0: basic or 64 Kbps digital channel).

that privacy and security of communications are guaranteed to the extent possible using the existing technologies, but as good as that of the PSTN networks (this may be difficult to achieve cost-effectively); and

e. An IP-based Intranet or VPN over the public Internet that can guarantee certain amount of bandwidth (e.g., 100 Kbps for G.711 coded voice signal without silence suppression) per admitted voice call with a prespecified amount of delay variation (e.g., less than 20 msec) and loss of packets (e.g., less that 3%).

**A Simple Call Flow**

Let us look at a very simple call setup scenario at a very high level where the LD call is routed over an IP network instead of a PSTN transport network. The CLASS-5 switch is providing a dial tone and other call access and delivery services to the phones at both the calling and called parties' premises.

The call control intelligence, which resides at the VoIP call server, receives the PSTN call setup messages via the SS7 SG or IP-PSTN MGW. When IMT-type links are used to connect the IP-PSTN MGW to the Intranet, call setup messages flow through the SS7 signaling gateway. When T1-PRI/CAS links are used to connect the IP-PSTN MGW to the Intranet, call setup messages flow through the same IP-PSTN MGW.

This ingress VoIP call server is also aware—via the system configuration—of the IP address of the ingress (call-originating) IP-PSTN MGW. It uses information from PSTN domain call setup messages—such as the initial address message (IAM, from the call-originating side)-type PSTN call setup message—to determine the E.164 addresses (telephone numbers) of the calling and called parties and to initiate a VoIP session between them using VoIP call control and signaling, as discussed in Chapters 2 and 3.

The ingress VoIP call server then uses H.225 (LRQ/LCF), SIP-T, or BICC messages—as discussed in Chapter 3—to determine the location of the egress VoIP call server. The egress VoIP call server returns the IP address of the IP-PSTN MGW, which can directly terminate the requested PSTN call. For the sake of simplicity, the ingress and egress VoIP call servers are shown in the same box in Figure 7-1.

At the same time, the egress CLASS-5 PSTN switch starts processing the incoming call setup request by capturing a two-way circuit and then checking for the availability of the called party by sending an "alerting" (for digital phone set) or "ring" (for analog phone set) message. The received response is the address complete message (ACM, a type of ISUP message [3]) that is received from the call-terminating side and is propagated to the call-originating side over (a) the SS7 network if the ingress, egress, and transport networks use PSTN or circuit-switching technologies or (b) the SS7 and IP networks if VoIP-based CLASS-4 or LD voice service is implemented. If the called telephone is not busy, the calling party hears the ring-back tone; otherwise, the called party is busy, and the calling party hears a busy tone. These tones are encapsulated over VoIP call control and signaling messages for transmission over the IP transport network (Intranet or VPN, as shown in Fig. 7-1).

If the called party is idle and answers the phone call (i.e., the handset goes off-hook), a "connection request" message is initiated from the egress side. This message is equivalent to the answer message (ANM, a type of call setup message) in the SS7 [3] network that initiates the billing process for the call. An RTP tunnel or session (see Chapter 2 and Reference 4 for details) is now established between the ingress and egress IP-PSTN MGWs by using the pre-specified RTP port numbers, as administered by the ingress and egress VoIP call servers. This RTP session runs over UDP/IP across the Intranet or VPN shown in Figure 7-1. The requested LD voice communication can now continue over this RTP session via appropriate mapping of the RTP session to the ingress and egress circuits, with the local access and delivery still using TDM or circuit-switch-based CLASS-5 networks.

As soon as the call is completed, either the caller or the called party goes on-hook, and the *disconnect* event sends the call release (REL, an ISUP message for call control [3]) message toward the other direction from the endpoint that initiated the on-hook action. A release complete (RLC, an ISUP message for call control [3]) now travels in the opposite direction—that is, toward the endpoint that initiated the on-hook action—to release the circuits on the access and delivery sides (both PSTN) of the network. The REL and RLC messages are translated into appropriate VoIP call control and signaling messages (e.g., BYE in SIP, Delete-Connection in MGCP, etc., as discussed in Chapter 3) to terminate the RTP session between the ingress and egress IP-PSTN MGWs in the IP-based transport network.

### Network Evolution Issues

The main advantage of VoIP-based LD service is that customers enjoy flat monthly rate–based billing for the calls within national boundaries. This is due to the fact that the voice sessions are transported over a distance-insensitive and shared IP-based network instead of over a circuit-switched PSTN transport network. Other advantages of VoIP-based LD service include (a) flexibility to customize the service per customers' requirements and (b) the ability to rapidly roll out new and emerging value-added services using server-based technologies. These advantages are enabling the Internet service providers (ISPs) and the competitive local exchange carriers (CLECs) to offer all-distance, IP-based voice or telephony services at discounted prices.

However, there are a few major issues that need to be addressed before VoIP-based LD service can achieve PSTN-grade quality, reliability, availability, and security. These include guaranteeing 99.999% of reliability and availability of services, consistently offering high-quality (e.g., toll grade or a MOS score of 4.0) voice transmission, and ensuring circuit-switch-type security of services.

As technologies improve, the IP network and related technologies will be able to support better availability, security, and quality of access, transmission, and delivery of voice traffic. These evolving technologies include one or more of the following:

a. Routing the packets for real-time voice sessions using an overprovisioned or overcapacity-based voice-grade transport network (e.g., one-way ETE delay of less than 100 msec, delay variation of less than 20 msec, and packet loss of less than 3%; for G.711, a coded voice signal without silence suppression with 20 msec of voice sample or packet);

b. Administration of voice call admission on the basis of ETE monitoring of multiplexing, storage, and bandwidth or transmission resources;

c. Categorization of real-time voice and loss-sensitive data into separate streams so that they can be multiplexed over different sets of RTP and UDP ports and, if required, can even be routed over different sets of IP addresses to guarantee the required quality of service; and

d. Deployment of IP version 6 (IPv6, IETF's RFC 2460/1883) or IP version 4 (IPv4, IETF's RFC 791) with IPSec infrastructure in the network.

Many next-generation network element manufacturers and service providers are exploring the effectiveness of these technologies for VoIP service in the pilot networks. These are discussed further later in this chapter and in Chapter 9.

It is well known [1–4] that PSTN networks are inherently more secure and reliable than VoIP networks and are capable of providing high-quality of transmission. However, they are neither open nor flexible enough to accommodate new value-added services as rapidly as VoIP-based networks.

Using the architecture shown in Figure 7-1, VoIP-based LD and other CLASS-4 services can be deployed as per the service capacity and capability requirements. For example, one can start with one VoIP call server, two IP-PSTN MGWs, a firewall and network address translator (NAT) device, and a VPN with a few call-originating and -terminating sites at the beginning. Then, as the demand increases, a network of VoIP call servers can be created, with each server controlling a cluster of IP-PSTN MGWs, and so on.

For large-scale deployment, service providers may consider using the architecture framework shown in Figure 1-9, as recommended by the Multiservice Switching Forum (MSF) in their recently published implementation agreements (available at www.msforum.org/techinfo/approved.shtml, 2001). The beauty of this architecture is that the functional elements used here are sufficiently modular or granular, and the interactions among these elements can occur over, IP links using various open or standard VoIP protocols such as SIP, MGCP, Megaco/H.248, and SCTP. These make the network architecture more scalable, growable, and proof of any emerging technologies. In addition, these characteristics can help the service providers launch new and emerging services—such as Internet call waiting, customized criteria-based call forwarding, instant messaging and conferencing, and so on—very quickly and economically.

## VoIP IN THE ACCESS OR LOCAL LOOP

In residential access networks, IP-based real-time voice or telephony service can be offered using a variety of access networking technologies. Recent developments in the technologies for access networking and physical transmission have significantly contributed to delivering broadband services to the home (BTTH, [5]). These include digital subscriber line (DSL, www.dslforum.org, www.dsllife.com, 2001) technologies, Ethernet in the first mile (EFM, www. efmalliance.org, 2001) technology, packet-cable and data over cable service interface specifications (DOCSIS, www.packetcable.com, 2001) technologies, and various WLL technologies. These are discussed in details in References 4 to 6.

In PSTNs, traditionally CLASS-5 switches along with twisted-pair copper wire–based local loops, are used to offer telephony service using channel associated signaling (CAS) [1,4,6]. Table 7-2 presents a list of the most widely used CLASS-5 features and services. A detailed list of all of the CLASS-5 features

**TABLE 7-2   Widely Used CLASS-5 Features and Services**

Automatic callback: automatic redialing of the last number dialed

Automatic recall: automatic dialing of the last incoming caller's phone number

Call blocking: blocking of certain outgoing calls by the subscriber

Call pickup: answering a call to one line from another line location by using an access code

Call transfer: transferring calls from one line to another

Call waiting: flashing of a text message (in the display of the phone set) or a background audio message/tone to announce a second incoming call

Call forwarding—busy line: forwarding incoming calls to another number when the dialed telephone is busy

Call forwarding—don't answer: forwarding incoming calls to another number when the call is not answered

Call forwarding—universal: unconditional forwarding of incoming calls to another number

Call forwarding—call-waiting calls: forwarding incoming call-waiting calls to another number

Call forward—remote activation: activation of call forwarding remotely from any other phone

Call hold: putting an active call on hold in order to pick it up from another line

Call intercept or anonymous caller rejection: intercepting or rejecting all incoming calls that block delivery of the caller's telephone number, name, or both

Caller's name and number (caller ID) delivery: displaying the calling party's telephone number and name after one ring

Called ID blocking: to blocking the calling party's identification (name, number, or both)

Cancel call waiting: special prefix code (e.g., *70) based dialing to cancel the call waiting feature for the duration of a call

Call tracing: activation of the incoming call tracing

Centrex features: the PSTN-hosted voice call processing feature used by business customers (discussed in Chapter 6)

Distinctive ringing: delivering different ring tones based on the number dialed over a single line

Extension bridging: programming one telephone number for multiple locations (requires the call forwarding and speed dial functions)

Make line/set busy: access code–based activation of "phone/line busy" appearance

Message waiting indication (MWI): a visual signal–based indication of the waiting messages (with display of time and date stamps)

Regulatory features: supporting the emergency dialing (911), directory assistance (411), CALEA, and other features and services

Selective call acceptance, forwarding, and rejection: preprogrammed lists–based acceptance, forwarding, and rejection of the incoming calls

Speed dialing: programming soft or hard buttons (using a one- or two-digit code) in the phone set for frequently dialed phone numbers

Teen services: caller ID and/or called number–based distinctive ringing, distinctive call waiting, directing the caller to a special mailbox, and so on

Three-way calling: conferencing with three callers

**TABLE 7-2** *(Continued)*

Utility telemetry service: utility companies' access to subscriber lines to receive utility usage data for billing purposes

Universal CLASS feature access: usage-based billing for access to the universally deployed CLASS features

Voice mail service: PSTN network host-based voice message recording by a caller and subsequent processing (retrieval, deleting, forwarding, archiving, etc.) by the called party

Wake-up call service: programming an incoming call at a prespecified time for wake-up service

Wide area telephone services (WATS): the capability that allows customers to make (OUTWATS) or receive (INWATS) LD calls and to have them billed on a bulk rather than an individual call basis

and services can be found in the corresponding generic requirements (GRs) that have been developed by Telcordia (www.saic.com/about/companies/telcordia. html, formerly Bellcore) for PSTN networks. Although more than 3000 CLASS-5 features have been developed, those presented in Table 7-2 are most useful and popular. Many of these services are CLASS-5 switch feature based, and some of them are SS7 network [3] based. For example, call waiting, call forwarding, three-way calling, and speed dialing are switch-based services, whereas automatic recall/call back, distinctive ringing, call trace, and selective call rejection are SS7 network-based services. In the next-generation PSTNs, the switch-based features and services may reside in the carrier-grade general-purpose computer servers with an IP interface, enabling the service providers to roll out new services quickly and economically. However, the VoIP-based telephony service may need higher bandwidth than that needed for circuit-switch-based telephony. For example, G.711-based voice coding needs 64 Kbps circuits for real-time voice communication over a circuit-switched (PSTN) network, whereas with the same G.711 coding, because of PPP/MAC/Ethernet, RTP, UDP and IP overheads, VoIP transmission needs more than 100 Kbps of bandwidth, as shown in Figure 2-2 of Chapter 2. In addition, very often, VoIP uses the same channel or pipe that is carrying non-real-time bursty packets for other services. Consequently, the challenges are to make VoIP-based telephony services as reliable, robust, bandwidth-efficient, and secure as those in the circuit-switch-based PSTN networks, without making network implementation and operation more expensive than that of PSTN.

The CATV or community antenna television (CATV, www.catv.org, 2001) network is another type of residential network that can be used to offer IP-based, real-time voice or telephony service. The newly allocated return path band (5 to 40 MHz) and the forward path band (600 to 750 MHz) [5,6] can be used to offer VoIP/IP telephony, as well as other advanced two-way video and entertainment (broadband) services. This can be achieved without jeopardizing the traditional TV and Internet data services that are delivered over other adjacent or nonoverlapping frequency bands. To support IP telephony, CATV network designers face the problems of designing circuitries for minimizing radio frequency (RF) interference and for splitting signals for real-time point-to-point multimedia services over the broadcast medium. Additionally, these networks must offer PSTN-grade reliability, security, QoS, customer service, and non-flat-rate billing when supporting the VoIP services. However, since the IP telephony service offered over the CATV network still remains unregulated, market penetration will probably increase significantly within the next 5 years.

The fixed WLL networks [5–7] can be used to offer IP telephony services as well. The technologies include (a) local multipoint distribution service (LMDS, which operates at a 27.5–29.5 GHz band and can cover a radius of up to 5 km), (b) multichannel multipoint distribution service (MMDS, which operates at a 2.15–2.70 GHz band and can cover a radius of up to 50 km), and (c) cable-less free-space optical transmission technology (operates at unlicensed hundreds of gigahertz to terahertz frequency bands and can cover a radius of up to 3 km). These technologies support high-bandwidth direct wireless channels to residential users to offer broadband data and TV services, and are very cost-effective in delivering communication services to sparsely populated and remote geographical areas. However, to support the VoIP service, WLL service providers face the same problems that the CATV service providers are facing. This is due to the fact that they operate at superhigh (gigahertz, terahertz) frequency bands. In addition, the WLL-based service providers must solve many of the well-known wireless transmission system (e.g., tower, cell, link) design problems in order to maintain PSTN-grade security, reliability, and QoS.

In this section, we describe high-level network architectures for the above three network scenarios that can be used to deploy VoIP-based CLASS-5 services to residential users.

**PSTN Networks**

The plain old telephone service (POTS) providers can use their existing twisted-pair copper wires (category 3 cables) to offer VoIP in the access network via DSL and EFM or IEEE P802.3ah (www.ieee802.org/3/efm, www.efmalliance. org, 2001) based access to the service.

The EFM technology is currently under development. EFM's goal is to support point-to-point connection over single-pair voice-grade twisted-pair copper wire and point-to-point and multipoint connections over optical fiber links. EFM is scheduled to be lab- and field-tested during 2003, with a plan for endorsement by the IEEE P802.3ah committee in late 2003. EFM will allow Ethernet frames to be directly transported over DSL, removing the need to use the ATM-based [4,6,8] layer 2 or link layer (as shown in Figure 2-10 of Chapter 2) transmission of information. Initially, EFM over copper (EFMC) wire will support a data rate of 10 Mbps over a distance of 0.75 km. Figure 7-2 shows EFMC for residential VoIP service in the access loop.

Transmission of voice over DSL can be achieved by using one of two sys-

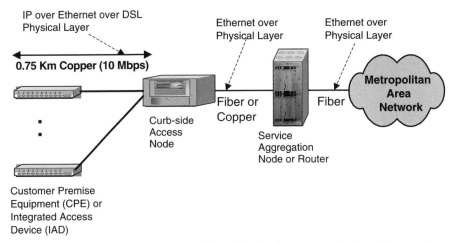

**Figure 7-2**  EFM (IEEE P802.3ah) with twisted-pair copper wire–based access from home.

tems. The first system uses ATM adaptation layer-2 (AAL2, ITU-T's I.363.2 Recommendation) based transport of TDM-formatted voice and call setup signaling, as suggested in ITU-T's I.366.2 recommendation-based voice service specifications developed by the ATM Forum and the DSL Forum (see, e.g., DSL Forum's Voice over DSL requirements specifications, TR-039, at www. dslforum.org/aboutdsl/Technical_Reports/TR-039.doc, 2001).

The other system uses ATM adaptation layer-5 (AAL5, ITU-T's I.363.5 recommendation) based transport for setup and control of voice call and transmission of voice signal using the protocols and methods discussed in Chapters 2 and 3.

The voice over AAL2/ATM option may be cost-effective from both service rollout and operations support viewpoints, but may not be as capable and flexible as the VoIP over AAL5/ATM option. Other advantages of using VoIP are the following:

a. Both software- and hardware-based IP phones can be directly (i.e., without using any adapter) connected to the customer premise equipment or integrated access device (CPE/IAD);

b. Many new and enhanced IP-based telephony services, such as instant conferencing and messaging and Internet gaming, can be introduced easily and quickly in a user-friendly (i.e., using a Web-based interface) fashion; and

c. Widely available and unified transport and networking (as shown in Figure 2-10 of Chapter 2) layer protocols such as TCP/UDP over IP can be exploited.

Current-generation DSL uses ATM cell (one cell is 53 bytes long, including header information) based layer-2 or link-layer transport, and this technology is mature enough for deployment in operational networks. Many companies are manufacturing DSL Forum and ITU-T specifications-compliant DSL modems embedded in the CPE/IAD and DSL access multiplexers (DSLAMs) that can aggregate several subscriber (CPE/IAD) lines into a single backhaul processing facility.

Although there are more than half a dozen variations of DSL, G.lite is most popular for asymmetric data rate (1.5 Mbps toward home and 640 Kbps from home) services to residential customers, and G.shdsl is most widely used for symmetric data rate (between 192 Kbps and 2.312 Mbps in each direction) services to business customers (see, e.g., www.dslforum.org/about_dsl.htm, 2001).

Since DSL uses the same twisted-pair copper wire used for telephone service, the DSL-based service can be expected to be as reliable, robust, and secure as the POTS service. However, for a variety of reasons, many local loops are not qualified to support DSL connections [5]. For example, if the physical length of the copper wire from the central office to the home is more than 5.0 km (3.0 mi), the signal strength will be significantly degraded. Similarly, customers who are served by remote cabinets or by digital loop carriers (DLCs) would not qualify for the DSL service. Also, if there is any loading coil or signal splitter in the twisted-pair copper line, DSL service cannot be offered over that line. Note that a loading coil is used to extend the reach of a loop so that telephone service can be offered to a remote location. A splitter filters the high-frequency signal for data service and the low-frequency signal for voice service, so the line is already being shared for voice and data services.

*An Architectural Option*  Figure 7-3 shows one possible architectural option for transmission of packetized voice over DSL in the access or local loop by deriving the call control and features from the CLASS-5 switch. The Telcordia (formerly Bellcore) defined GR-303 or the TR-008 link (in the United States) or ITU-T's V5.1 or V5.2 link (in Europe) can be used to connect the DSLAM to the CLASS-5 switch via a voice gateway. The DSLAM aggregates traffic from multiple CPEs/IADs and delivers them to the ATM access network, which switches the voice traffic to the voice GW and the data (or Internet) traffic to the ISP's IP network. The voice GW receives the AAL2/ATM-encoded voice and call signaling cells (53 byte packets) and removes the encapsulation headers from the traffic so that it can delver a steady stream of TDM-formatted information over the GR-303/TR-008 or V5.1/V5.2 interface to the CLASS-5 switch.

In this scenario, the service providers can use the existing CLASS-5 switch and associated facilities for call control, billing, and feature delivery. Therefore, this option protects the investments in the call control and features delivery modules of the existing traditional CLASS-5 switch. It also allows the service providers to roll out packetized voice service on a small scale before rolling out

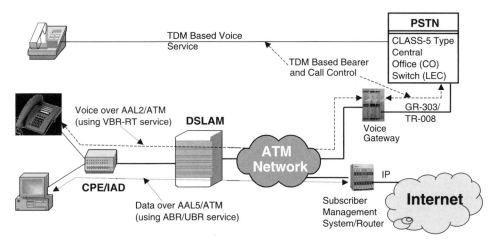

**Figure 7-3** An architecture for implementing Voice over DSL using GR-303/ TR-008 (in the United States) based voice GW (AAL2/5: ATM adaptation layer 2/5; ABR/UBR: available or unspecified bit rate; CPE/IAD: customer premise equipment/ integrated access device; DSLAM: digital subscriber line access multiplexer; TDM: time division multiplexed or circuit-switched line; VBR-RT: variable bit rate—real-time).

an IP-based infrastructure (as shown in Figure 7-4 and discussed later) for both call control and media (voice traffic) transmission.

However, additional circuits, network elements such as DSLAMs, ATM access networks and switches, voice GW with GR-303/TR-008 or V5.1/V5.2 and ATM interfaces, and QoS provisioning and management facilities are needed on the service providers' premises to deliver integrated services to customers. The CPE/IAD on the customer's premises contains an embedded DSL modem, and in some cases may contain firewalls (to support the security requirements) and routers (for in-home networking) as well. The CPE/IAD usually supports multiple telephone lines over RJ-11 interfaces and two or more Ethernet (RJ-45) interfaces for data services. For POTS calls, media and signaling information (i.e., voice, features), transmission can occur over QoS-guaranteed AAL2/ATM circuits (Fig. 7-3) using variable-bit-rate real-time (VBR-RT) service [4,6,8]. Data or Internet services can be offered via AAL5/ ATM circuits using available or unspecified bit rate (ABR, UBR, [4,6,8]) type service, as shown in Figure 7-3.

*An Alternative Architectural Option*   Figure 7-4 shows another architecture in which call control and advanced call-related services and features are delivered not from the CLASS-5 switch but from the IP-based call controller (CC) and feature servers. This architecture also uses the IP-PSTN MGW, the CC and MGC, and the SS7 signaling gateway to facilitate interactions with PSTN transmission (TDM circuits) and signaling (SS7) networks. This option is the

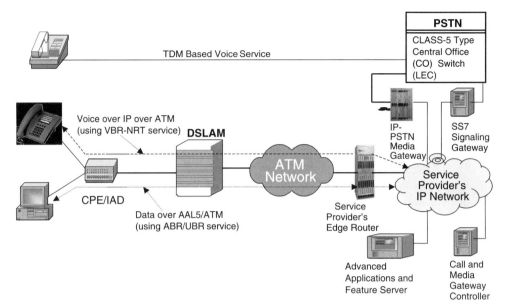

**Figure 7-4**   An Architecture for implementing Voice over IP over DSL using the IP-PSTN Media Gateway, SS7 signaling gateway, call and media gateway controller, and applications and feature server (AAL5: ATM Adaptation Layer 5; ABR/UBR: available or unspecified bit rate; CPE/IAD: customer premise equipment/integrated access device; DSLAM: digital subscriber line access multiplexer; TDM: time division multiplexed or circuit-switched line; VBR-NRT: variable bit rate—non-real-time).

most suitable one for IP network–based service providers who want to enter the VoIP service market.

As shown in Figure 7-4, the IAD and the IP-PSTN MGW are configured as the clients of the call and MGC and are controlled via MGCP or Megaco/H.248 protocols (these protocols are discussed in Chapter 3).

The IAD supports many types of end devices, such as, two or more traditional POTS phones via RJ-11 jacks, SIP phones, and other IP-enabled devices such as PCs over RJ-45 jacks. The IP-PSTN MGW adapts media information (i.e., voice traffic) from a TDM-formatted signal to IP packets for delivery over the service provider's IP network for an ETE RTP (over UDP over IP) session specific to a voice call.

The SS7 signaling GW accepts the VoIP call setup and control messages from the call and MGC, and translates them into appropriate ISUP and TCAP messages [3] for delivery them over SS7 A-links (or another type of link) to the SS7 network.

The applications and feature server host the services and features associated with a call and deliver them to the calling parties. As new applications and services—such as instant conferencing and messaging and Internet gaming—

are developed, these can be introduced to the customers just by adding complementary servers to the applications and services complex. This enables the service providers to roll out new services rapidly and very cheaply. Note that in traditional PSTN networks, introduction of a new service may take more than a year for software and system integration because of lack of openness and modularity of the service or system architecture.

In the networking scenario shown in Figure 7-4, an on-net call may not need to use any resources from the traditional PSTN network, and for off-net calls, it may be necessary to use only a minimal amount of resources for call setup and control. This is due to the fact that the on-net calls may originate from and terminate at either the same or any two call and MGW controllers connected to the same (service provider's) IP network. The IP-PSTN MGW would not be involved in the RTP tunnel or session that is needed for transmission of a voice signal during such a call.

## CATV Networks

As mentioned earlier in this chapter, the newly allocated upstream (below 40 MHz) and downstream (600 to 750 MHz) bands can be used for real-time two-way and interactive services including the VoIP service. However, the development of equipment and market trial of the VoIP service over CATV networks did not start until the DOCSIS 1.1 (data over cable service interface specifications version 1.1, www.cablelabs.com, 2001) based packet-cable architecture framework 1.0 (www.packetcable.com, 2001) was defined for delivering IP-based services to CATV users.

DOCSIS 1.1 defines how to provide data services, such as Internet services, over hybrid fiber coax (HFC)–based cable TV networks by incorporating a cable modem (CM) and a multimedia terminal adapter (MTA) on the customer's (or client's) side, and by adding a cable modem termination system (CMTS) at the head end. The CM performs basic RF modulation/demodulation, and signals and controls the services offered over the media streams. In addition to supporting interfaces and signaling for TV services, the MTA contains CODECs, digital signal processing (DSP) chipsets, and interfaces such as RJ-11, RJ-45, and USB. These features are required to support all of the signaling and encapsulation functions for delivering the PSTN CLASS features and the necessary level of QoS for supporting the VoIP service. The CMTS is an HFC network head-end device that provides connectivity to the wide-area IP network. CMTS implements DOCSIS' RF interface specifications-based media access control for managing the connection and services to the MTA via the CM. The services include QoS signaling, traffic shaping, packet marking, reserving bandwidth, supporting data and terminal security via encryption, key management and CM authentication, and so on.

Packet-cable architecture framework 1.0 defines the network elements and their interfaces that are required to interconnect DOCSIS 1.1 HFC network's CMTS to the PSTN network via a managed IP backbone or transport

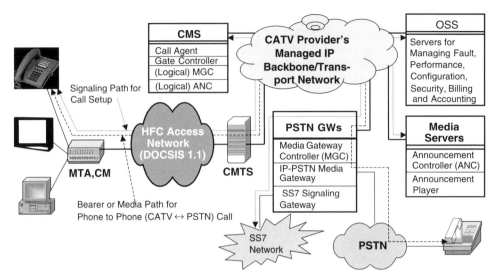

**Figure 7-5** VoIP service delivery using the CATV network using DOCIS 1.1 and PacketCable 1.x standards (CM: cable modem; CMS: call management server; CMTS: cable modem termination system; GWs: gateways; MTA: multimedia terminal adapter; OSS: operations systems support; PSTN: public switched telephone network; SS7: signaling system No. 7).

network to support both on-net and off-net VoIP-based telephone service. These specifications are available at the packet-cable website (pkt-tr-arch-v01-991201.pdf and PKT-TR-ARCH1.2-V01-001229.pdf at www.packetcable.com/ specifications/, 2001). The network elements required to support the VoIP service are (a) PSTN gateways, (b) announcements or media servers, (c) a call management server, and (d) operations systems support servers, as shown in Figure 7-5.

The PSTN gateways include an MGC, an IP-PSTN MGW, and an SS7 SG. The MGC controls the IP-PSTN MGs and both on-net and off-net (i.e., the called party is a PSTN-hosted phone) call setup requests. It makes the call routing decisions, maintains the call states, and ensures the authenticity of the entities communicating with it. The IP-PSTN MGW provides media (or bearer) connectivity between the PSTN network and the packet cable's managed IP backbone or transport network, and performs the necessary signaling, transcoding, and echo cancellation functions to guarantee high-quality continuity of media adaptation. The SS7 SG mediates secure exchange of ISUP (for call setup) and TCAP (for call feature/applications) messages between PSTN's SS7 [3] network and the MGC or the call agent (within the call management server).

The announcements or media servers consist of an announcement controller and an announcement player. The announcement player is a media resource server; it receives the input trigger (e.g., one or more DTMF digits) from the MTA and plays a media file under the instruction of the announcement controller. The announcement controller monitors the state of the VoIP calls from the call management server (discussed below) and manages play-out of any announcement during the VoIP session.

The call management server contains a call agent and a gate controller. It also logically contains an MGC (defined above) and an announcement controller (defined above) via proxy arrangements, for example. It provides call control and signaling associated services to MTA, CMTS, and PSTN gateways. For example, the call agent can use MGCP-based signaling (discussed in Chapter 3) to provide call progress tones, digit collection, delivering CLASS, LNP, directory assistance, and any other pre- and postcall setup features to the MTA (PKT-SP-EC-MGCP-I04-011221.pdf at www.packetcable.com/specifications/, 2001). The gate controller dynamically manages the quality of service of the VoIP calls using the method developed in the packet cable dynamic QoS specifications (PKT-SP-DQOS-I03-020116.pdf at www.packetcable.com/specifications/, 2001).

The operations systems support consists of a set of servers to support service management, network element provisioning and configuration management, CDR maintenance, distribution of keys to support security, and so on. These include management of network configuration and faults, performance of the system, security of the services, and accounting and billing for the services rendered to the customers, as per ITU-T's TMN framework/specifications.

Packet cable supports both network-based and distributed call signaling. In network-based call signaling (e.g., as in MGCP, discussed in Chapter 3), a client server (master-slave or hierarchical) based model is used. It is assumed that all of the call control intelligence and features reside in the server, with the clients (the MTA and CM) being dumb peripherals. In this scenario, the emerging call control and service features can be introduced to the customers very easily by upgrading the software, hardware, or both on the server side. Using the PSTN network design paradigm, the scalability and reliability requirements of the services can be fulfilled.

In distributed call signaling (e.g., as in the SIP, discussed in Chapter 3), a peer-to-peer (or flat) model is used for call control and feature delivery, and the endpoints (the MTA and CM) are assumed to be smart devices. In order to use the emerging features and services, the software and hardware of the endpoints must remain up-to-date. In this case, the Internet design paradigm can be used to satisfy the scalability and reliability requirements of the services.

A simple generic call flow for a call from a CATV network-attached phone to a PSTN-attached phone is also shown in Figure 7-5. The call agent inside the call management server controls the phone attached to the MTA-CM and hence provides a dial tone, collects the dialed digits, facilitates play-out of announcements, identifies the called and calling parties, and sends the call setup request via the SS7 signaling GW to the SS7 network. The IP-PSTN MGW sets up the bearer path (shown by the dashed line in Figure 7-5) under the instructions of the call agent and the MGC. The CMTS and the servers in the operations systems support complex ensure maintenance of security and voice path quality and perform accounting functions for the call.

Although many CATV service providers are rolling out VoIP-based telephony services to their customers, many problems—such as providing uniform QoS and billing when a call passes through several service providers' networks, supporting PSTN-grade reliability, availability, and regulatory features even during failure of the supply of electricity, and so on—remain to be solved. The packet cable standardization group is working to solve these problems; for example, DOCSIS 2.0 specifications for symmetric transmission (an upstream data rate of up to 30 Mbps) of a variety of media streams including packetized voice, video, and data are currently being developed, and packet cable specifications 1.1 and 1.2 for various signaling and PSTN-grade services are either becoming available or are expected to be ratified in 2003.

### Broadband Wireless Access (Local Loop) Networks

As mentioned earlier, fixed WLL technologies can be used to deliver broadband services including real-time IP telephony services.

The LMDS (local multipoint distribution service, 27.5–29.5 GHz band) and free-space optical transmission links (lasers operating at bands of hundreds of gigahertz to terahertz) can be used either to replace the local telephone access loop or to offer complementary services to intra-LATA customers, because these technologies can offer services within 3 to 5 km of the base transmission station. LMDS can support a traffic rate of up to 1.50 Gbps downstream (to users or homes) and up to 200 Mbps upstream (from users).

The MMDS (multichannel multipoint distribution service (2.15–2.70 GHz band) can provide service over a zone which spreads over a 50 km radius of the base transmission station, and can be used to offer both intra- and inter-LATA telephony and other supplementary services.

The IEEE 802.16 Working Group (www.ieee802.org/16/, 2001) on Broadband Wireless Access Standards [7] has been working to create a new connection-oriented media access control (MAC) procedure that can not only carry arbitrary protocols transparently but also support QoS requirements for very-high-bit-rate wireless access to metropolitan area networks (MANs). Using an authenticated key management protocol, a privacy sublayer is also being defined to provide encryption and protection of services from unauthorized uses. The wireless-MAN air interface specifications assume a point-to-multipoint topology, with a controlling base station connecting residential and business customers to various public networks. Although the current focus of the IEEE 802.16 committee is on the MAC protocol for the frequency spectrum in the 11.0 to 66.0 GHz band, in the future they will consider the frequency band between 2.0 and 11.0 GHz as well [7].

Many factors influence the numbers of subscribers per wireless cell that is being served by one or more base stations and the amount of bandwidth each subscriber can get in each direction (to/from home). These include the size of the spectrum (frequency band); the signal coding technique used, which determines the number of bits packed per hertz of the uplink and downlink (to home) channels; the user multiplexing method implemented at the base station; and the capabilities of the antennae.

As mentioned earlier, to deliver voice over IP to residential and corporate users, the WLL-based service providers must find cost-effective solutions to the challenges of designing and operating super-high-frequency (i.e., gigahertz and terahertz bands) wireless networks that can offer access services to both line-of-sight and non-line-of-sight receivers. The challenges consist of ensuring (a) availability, reliability, and security of transmission, (b) proper strength of the signal even in the presence of interference, fading, failure of the electric power supply, and adverse atmospheric conditions, and (c) appropriate utilization of the frequency spectrum for optimizing cost and channel capacity constraints without affecting the QoS [7,9].

A handful of companies, including Beamreach (www.beamreachnetworks. com, 2001) and Terabeam (www.terabeam.com, 2001), have developed field-deployable base stations, antennae, and receiver devices for WLL-based high-bit-rate data services and real-time services such as video on demand, voice over IP, and so on. Some corporations and service providers are either field-testing or have deployed these technologies in prototype networks.

## IP-BASED CENTREX AND PBX SERVICES

As discussed in Chapter 6, centrex services can be offered to small and medium-sized corporate customers by replacing the T1 links (from the service provider's central office) to the customer's premises by DSL or other high-speed IP links (e.g., a gigabit-Ethernet link). On their premises, the customers need to maintain a local POTS access GW if they want to continue using the POTS phones and a LAN that is capable of supporting both real-time and non-real-time data communications services. The same LAN can now support both data terminals and IP phones, with signaling and call control functions being derived from the centrex feature GW on the service provider's premises, as shown in Figure 7-6. Although either ITU-T's H.323 or IETF's SIP protocol (discussed in Chapter 3) can be used for VoIP call control and signaling, because of its low complexity, higher capability, and flexibility in adding new features, it may be more practical to use SIP-based call control and signaling. In addition to saving operational costs, the IP-based centrex supports many new services such as IP-based virtual private networking (IP-VPN), unified messaging, Web-based configuration management, and viewing and payment of monthly bills.

Large enterprises or corporations usually have their own IT (information technology) service maintenance departments and probably use their own PBX systems for intra- and intersite voice communications services. In order to incorporate IP telephony in a traditional PBX, either a VoIP line card or

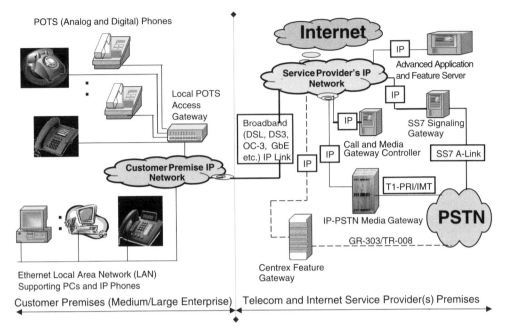

**Figure 7-6** Evolution of traditional centrex service offering to IP and technologies-based centrex service delivery. The connections shown by the dashed line are required when PSTN call and feature control reside in the PSTN network, and (a) the SS7 signaling gateway, call and media gateway controller, and (b) advanced feature server are not deployed. The centrex feature gateway supports the GR-303/TR-008 interface to PSTN and may contain the VoIP call controller and media gateway. (Source: Fig. 6-2)

an adjacent IP telephony GW, along with a VoIP call controller, must be added.

The VoIP GW and the CC are interconnected to the same LAN that provides IP/Ethernet-based data transmission services to the users, and the corporate IT department manages this LAN. The LAN in one corporate site can be connected to the LANs in other corporate sites using wide area IP links. Thus, the costs for maintaining T1 connectivity to the PSTN service provider's central office—for telephony service—can be significantly reduced, if not entirely eliminated. Direct PSTN connectivity over T1 link(s) can be maintained only to the IP-PBX at the corporate headquarters.

Deployment of the VoIP GW also ensures that the investments in the existing circuit-switch-based PBX and phone sets are protected until they fully depreciate.

The VoIP CC can be deployed at one or more corporate sites but not at all of them. Since all of the sites are interconnected over the corporate wide area IP network, the calls from traveling corporate users, and from small or remote corporate sites, can be controlled by the VoIP CC at headquarters. This not only guarantees security and QoS, but also helps maintain centralized accounting, billing, and auditing functions. These issues are discussed in detail in Chapter 6.


## EPILOGUE

VoIP technology, as described in Chapters 2 and 3, is currently deployable for telephone services in both public and enterprise or corporate networks. For traditional LD or CLASS-4 services, the deployment of VoIP has fueled competition and pulled prices down. Furthermore, because of deregulation of the U.S. telecom industry (www.fcc.gov/telecom.html, 2001), both competitive and incumbent telephone service providers (local and LD) are introducing many new VoIP and Web interface–based advanced services. These include bundling services such as unified messaging, instant multimedia conferencing, and find-me/follow-me service with local and LD phone services for nominal fees.

The emerging competitive service providers include the traditional CATV and broadband wireless access service providers. Using VoIP in the local loop over DSL, cable channels, and WLL channels (the Federal Communications Commission [FCC] manages the spectrum for some of these channels, as mentioned at www.fcc.gov/oet/), these service providers are packaging Internet service, traditional and advanced (digital) TV and other entertainment services, and IP-based LD voice service for unified billing and customer care.

The traditional PSTN service providers are also upgrading their network infrastructures to support (a) DSL-based local and LD VoIP and Internet services to residential customers and (b) IP centrex and DSL or other high-speed IP link-based intercorporate-site connectivity of IP-PBXs for enterprise or corporate customers.

However, as discussed in the Epilogue of Chapter 6, many network planning, deployment, regulatory, QoS, and customer service–related issues still need careful considerations before widespread deployment of VoIP can occur. This is due to the fact that the public networks usually provide a heterogeneous mix of services to hundreds of thousands of customers over a wide geographic area; hence, the technology (e.g., VoIP) they choose must be sufficiently reliable and scalable to guarantee cost-effective deployment of the service. For example, if a multilayer server- and GW-based architecture—as shown on the left side of Figure 1-8 in Chapter 1—is chosen for the VoIP service, network elements such as IP-PSTN MGWs, call managers, SGs, and so on must be able to provide nonstop service to hundreds of thousands of customers without any degradation of availability and response time even when a large number of customers simultaneously request the same set of services.

*Continuous availability* of service can be guaranteed by using a combination of (a) component- and interconnection-level redundancies within the network element and (b) distributed interconnection and operation of the network elements, as in the Internet. Note that because of the distributed nature of implementation and operation of IP-based networks, any outage of a network link, router, or any other network element can be detected within tens of milliseconds (see, e.g., www.internet2.org, 2001). Consequently, the traffic can be redirected to a healthy segment of the network within that period of time. This soft handling of faults may increase the call setup time, and the talking parties may hear a momentary silence or glitch rather than rejection of the call setup request and dropping of already established calls, as may happen in the traditional PSTN networks. This feature of IP-based telephony service may even be beneficial to the users during massive network faults or disaster in a certain geographical area.

The network elements that are used to deploy the VoIP service must support Web-based management, open/standard network interfaces, protocols, and APIs so that service capacity can be added easily and new features can be introduced rapidly by upgrading the codes of the network elements in different layers (Fig. 1-8). For example, to support *authorized electronic surveillance* of network elements, communication links, and services, additional servers can be added in the signaling and control or applications and services layer (Fig. 1-8). This ensures that the emerging service providers comply with the Communications Assistance for Law Enforcement Act (CALEA, www.fcc. gov/calea/) requirements even when they are using IP-based network elements for telephony service. The proposed interim electronic surveillance specification for VoIP-based telephony services over CATV network (pkt-sp-esp-I01-991229.pdf, at www.packetcable.com/specifications/, 2001) recommends such a method to support CALEA. Similarly, to support emergency public safety services, such as the E911 service (www.fcc.gov/911/), additional GWs and servers can be introduced in the signaling and control layer (Fig. 1-8) to route a 911 call from an IP or POTS phone to the designated public safety answering point (PSAP), along with the caller's phone number and location information.

In order to satisfy the *regulatory requirements* of the Industry Analysis and Technology Division of the FCC (www.fcc.gov/wcb/iatd/), the traditional PSTN-based wire-line telephone service providers must guarantee continuous availability of telephone service to residential customers. Continuous blocking of calls and noticeable outage in the delivery of a dial tone to residential phones must be captured and reported to the FCC in the Telco service quality reports (www.fcc.gov/wcb/asd/sq.html). To provide continuous availability of services, the traditional telephone service providers use not only backup electricity generators and battery plants, but also primary and secondary central office (CO) switches, and deploy alternate or backup links to connect CO switches within a LATA and between LATAs. Thus, the physical line from the main distribution frame (MDF) to homes remains the only unprotected facility for delivering telephone service to homes. To replicate this type of service availability, emerging service providers are doing one or more of the following: (a) using rechargeable battery–based service and power backup in telephone sets and deploying UPS devices (see, e.g., www.apc.com/products/, 2001) to absorb short-term surges and disruptions of the power supply to customer premise equipment (CPE) such as modems, routers, and switches; (b) utilizing the unused wires in the Ethernet cable to deliver power from the EtherSwitch to telephone sets; and (c) using special heartbeat-type signals from the CPE so that customer service representatives can advise customers about the status of the primary and backup power supplies in their devices.

In order to solve *authentication and information (signaling and media) security* problems, session-based verification of user identity and large-key-based information encryption can be used. These call for the use of authentication servers, firewalls, network address translator (NAT) and proxy devices, and key distribution servers within the signaling and call control layer (Fig. 1-8) or the operations system support layer of the network architecture, and these servers can provide a framework to support IPSec in the IPv4-based networking environment. Note that packet cable's interim security specifications (PKT-SP-SEC-I05-020116.pdf at www.packetcable.com/specifications/, 2001) and architecture recommendations (pkt-tr-arch-v01-991201.pdf and PKT-TR-ARCH1.2-V01-001229.pdf at www.packetcable.com/specifications/, 2001) also suggest the use of this type of framework to support security and authentication of VoIP service over CATV networks.

To maintain *PSTN-grade quality of voice transmission and call setup* in multiservice IP networks supporting both real-time voice transmission and non-real-time data services, real-time traffic (such as voice packets) must be identified and marked right from the source. This guarantees that these packets are dispatched at higher priority in all layers (as shown in Fig. 2-10 of Chapter 2) within the network. Then, in order to maintain an acceptable level of voice quality (e.g., a MOS of 4.0), the access, transport, and delivery networks must be designed in such a way that the one-way ETE delay of 150 msec (as per ITU-T's G.114 specifications [2]) can be guaranteed. Packet cable's interim dynamic QoS specifications (PKT-SP-DQOS-I03-020116.pdf and pkt-sp-iqos-i01-001128.pdf at www.packetcable.com/specifications/, 2001) recommend a framework in which ETE QoS for a VoIP session is maintained by concatenating QoS-guaranteed access, transport, and delivery segments. A combination of call admission control (e.g., using common open policy service, or COPS) and application, network, and link layer QoS signaling mechanisms (e.g., use of the SDP parameters during call setup, RSVP signaling, rate-guaranteed packet emission, etc.) is used to achieve the desired quality of call establishment and voice transmission during the conversation. Deployment IP version 6 (IPv6, IETF's RFC 2460/1883, www.ipv6.org, www.internet2.org, 2001) based addressing, and other security and QoS offerings in the network, can satisfy many authentication, security, and QoS requirements. However, since IPv6 is not yet widely deployed, this may create many service incompatibility or unavailability problems.

Note that the initial costs for deploying a multilayer server- and GW-based architecture (Fig. 1-8) for VoIP service may be higher than that of traditional PSTN-based implementation of POTS. However, even then, because of supplementary benefits such as (a) convergence of network managements and infrastructures, (b) openness of the interfaces and protocols used in the network elements, and (c) flexibility to add new services and features easily and quickly, many medium-sized and large enterprises, as well as emerging carriers and service providers, are deploying this type of architecture for IP-based telephony services. They are devising cost-effective and innovative solutions to many QoS, security, reliability, and availability problems and deploying them in their networks. Once these innovative solutions reach maturity, they will be sufficiently stable so that they can be standardized for wide-scale deployment in the next-generation public networks. VPN is one such technology that is being widely used for interconnecting the IP-PBXs at different geographical sites of a corporation [10]. VPN technology can be extended to create both intra-LATA and inter-LATA virtual trunk networks (VTNs) using the same IP networking methods.

These new technologies will make the deployment of VoIP and IP telephony as realistic as PSTN-based POTS in both corporate and public networks. Many organizations are working to achieve that goal, with active participation from service providers, equipment manufacturers, and regulatory bodies. These include the Internet2 consortium (www.internet2.org, www.internet2.edu, 2001), Multiservice Switching Forum (www.msforum.org, 2001), Alliance for Telecommunication Solutions' Signaling for VoIP (SVoIP) effort (www.atis.org, 2001), and International Softswitch Consortium (www.softswitch.org, 2001).

## REFERENCES

1. W. J. Goralski, SONET, Second Edition, McGraw-Hill Book Companies, New York, 2000.

2. G.114 Recommendation, One-way Transmission Time, ITU-T Geneva, 1996.

3. T. Russell, Signaling System #7, Second Edition, McGraw-Hill Book Companies, New York, 1998.

4. W. Stallings, Data and Computer Communications, Sixth Edition, Prentice-Hall, Upper Saddle River, New Jersey, 2000.

5. B. Khasnabish, "Broadband To The Home (BTTH): Architectures, Access Methods and the Appetite for It," IEEE Network, Vol. 11, No. 1, pp. 58–69, January/February 1997.

6. M. Tatipamula and B. Khasnabish, Editors, Multimedia Communications Networks: Technologies and Services, Artech House Publishers, Boston, 1998.

7. C. Eklund, R. B. Marks, K. L. Stanwood, and S. Wang, "IEEE Standard 802.16: A Technical Overview of the WirelessMAN™ Air Interface for Broadband Wireless Access," IEEE Communications Magazine, Vol. 40, No. 6, pp. 98–107, June 2002.

8. ATM Forum, ATM Traffic Management Specifications, Version 4.0, 1996.

9. Y.-B. Lin, B. Khasnabish, and I. Chlamtac, "The Wireless Segment of Enterprise Networking," IEEE Network, Vol. 12, No. 4, pp. 50–55, July/August 1998.

10. B. Khasnabish, "Next-Generation Corporate Networks," IEEE IT Pro Magazine, Vol. 2, No. 1, pp. 56–60, January/February 2000.

# 8

# VoIP FOR GLOBAL COMMUNICATIONS[1]

This Chapter discusses how IP-based voice communications can be deployed for global communications in multinational enterprises and for international calling by residential PSTN customers. In traditional PSTN networks, various countries use their own version of the ITU-T standards for signaling or for bearer or information transmission. When IP-based networks, protocols, interfaces, and terminals (PCs, IP phones, Web clients, etc.) are used, unification of transmission, signaling, management, and interfaces can be easily achieved. We discuss a possible hierarchical architecture for controlling IP-based global communications in a hypothetical multinational organization.

## VoIP IN MULTINATIONAL CORPORATE NETWORKS

Large multinational companies with global operations usually manage multiple network infrastructures for voice and data services. For data networking they commonly use IP, frame relay (FR), asynchronous transfer mode (ATM), and other networking technologies [1]. For voice communications—depending on the number of employees in a location—they either deploy PBX or use centrex services from telecoms local with, for example, T1-based (24 DS0 lines over a $24 \times 64 = 1.536$ Mbps line) PSTN connectivity in North America, E1-based (32 DS0 lines over a $32 \times 64 = 2.048$ Mbps line) PSTN connectivity in Europe, and so on [1].

---

[1] The ideas and viewpoints presented here belong solely to Bhumip Khasnabish, Massachusetts, USA.

**117**

If PSTN centrex-based services are used for voice calls, the costs for service from the telecom may be very high but the on-site maintenance costs will be low. If PBXs are deployed, two different network infrastructures must be maintained—one for data services and the other for voice services—in every corporate location. This involves two different sets of monthly bills and two different sets of personnel for maintenance and procurement of network elements such as phones, PBX line cards, routers, switches, UPS, and so on. By using IP-PBXs and consolidating these two network infrastructures into a single IP-based network infrastructure, multinational corporations can reduce operational expenses, including the expenses related to voice calls, and can introduce advanced productivity-enhancing services very quickly and economically, as discussed in detail in Chapter 6.

If circuit-switch or PSTN networking technologies are used to interconnect the PBXs in different countries, multinational corporations have to find a PSTN service provider who offers call signaling (including translation) and media transmission services internationally. Note that for every three E1 links terminating at a site in Europe, a corporation may need to deploy at least four T1 links in a North American site. This arrangement is expensive, although it may provide the flexibility to dial the phones in international locations by using a one- or two-digit prefix and a five- or seven-digit phone number instead of using country code, city code, and phone number–based dialing.

By deploying IP-based PBXs and interconnecting them using intercountry IP links with a guarantee of availability, reliability, security, and performance, flexibility of dialing and cost savings can be achieved simultaneously. A network of these widely available intercountry IP links can support high-quality transmission, and can create a global VPN that can be used for voice and data communications within the corporation across multiple distant LANs.

Figure 8-1a shows the migration of PBX-based telecommunications to an IP-PBX-based infrastructure in a North American site of a multinational corporation. Figure 8-1b shows the migration of PBX-based telecommunications to an IP-PBX-based infrastructure at a site in Europe of a multinational corporation. Note that the telephone sets, their interfaces, and the PSTN-side trunks are different in Europe and North America, but the IP phones, their interfaces, and the IP links are the same all over the world.

Figure 8-2 presents an overall hierarchical architecture for introducing VoIP service globally using IP-based network. The IP-PSTN MGWs of Figure 3-8 are now replaced by the IP-PBXs. The VoIP GW and CC (Fig. 8-2) control the resources in the VoIP line cards of the IP-PBX and route the intersite telephone calls over the IP-based network.

Note that along with the network elements required to support the VoIP service, the local variants (e.g., North American, European, Japanese) of PSTN or circuit-switched equipment (e.g., PBX, phones) and wiring can also be maintained until they fully depreciate. This strategy provides a graceful transition to an IP-based converged network for both voice and data services. As described in Chapter 6, the additional network elements required to support

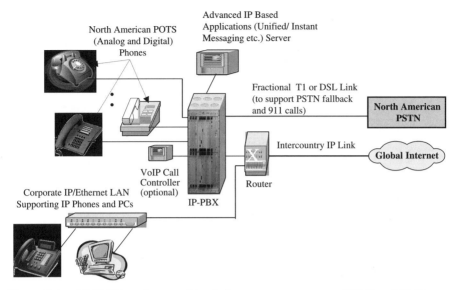

**Figure 8-1a**  IP-PBX-based networking infrastructure to support POTS and VoIP service simultaneously in a North American location (e.g., Boston, Massachusetts) of a multinational corporation.

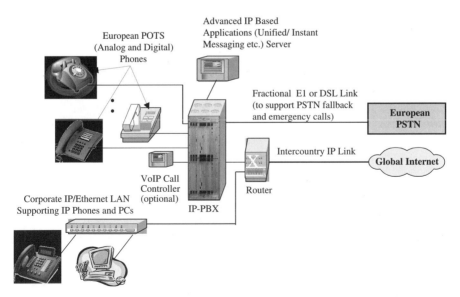

**Figure 8-1b**  IP-PBX-based networking infrastructure to support POTS and VoIP service simultaneously in a location in Europe (e.g., Paris, France) of a multinational corporation.

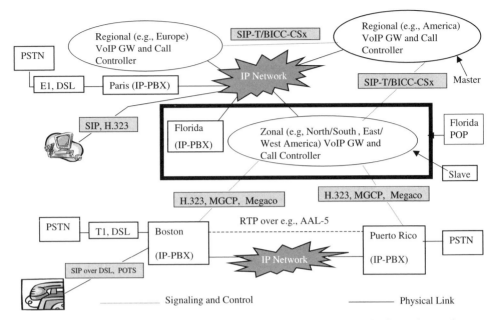

**Figure 8-2** An architecture for a packet-based global network for advanced or enhanced VoIP and POTS services in a multinational corporation.) (Source: Adapted from Fig. 3-8)

the VoIP or IP-telephony service within a corporation are IP-PSTN MGWs or VoIP GWs, VoIP call servers or call managers, IP phones, an uninterrupted power supply (UPS), and Ethernet and IP switches and routers capable of supporting the QoS needed for transmission of packetized voice signals in real time. In addition, when IP version 4 (IPv4)–based addressing is used, network elements such as firewalls, authentication and key distribution servers, a network address translator (NAT), and so on are also required to resolve many of the security and authentication problems that corporations are facing today while trying to use IP for voice communications. Alternatively, IP version 6 (IPv6)–based addressing can be deployed, which supports many of the required QoS, service, and security and user authentication features. The scalability of the selected networking technique and the service architecture must also be carefully analyzed before deployment; these will guarantee that the installed techniques and architectures satisfy the projected growth requirements of network and service infrastructures.

As mentioned earlier, using intercountry IP links, the IP-PBXs in international corporate locations can be interconnected, and a network of these IP links can create a global IP-VPN for the corporation. Traditional service level agreement (SLA) parameters for VPNs include availability of bandwidth and reliability of the link, including mean time to respond and mean time to repair during service outage. However, if the same VPN is used for real-time voice communications, significant attention must be given to the additional short-term (i.e., calculated over a short time interval) performance parameters such as one-way end-to-end (ETE) latency or delay, variation of delay or delay jitter, and percentage of packets lost, as discussed in IETF's RFCs (RFC 2475 and RFC 3198) and in Chapters 4, 6, and 7. The *short* time interval is equivalent to the length of a typical real-time voice conversation or session, which could be 3 to 5 min or longer. The short-term performance parameters not only determine the availability of a dial tone and the amount of time it takes to establish a voice call, they also drastically influence the quality of voice signal transmission during a conversation, as discussed in Chapter 4 in the context of QoS requirements and in Chapter 6 in the context of NGENs. For example, if G.711- or PCM-based voice coding—which produces a 64 Kbps bit stream—is used with a voice sample or packet size of 20 msec, an RTP session bandwidth of more than 100 Kbps is required (as shown in Fig. 2-2 of Chapter 2), with no *more* than 150 msec of one-way ETE (or mouth-to-ear) delay [2], approximately 20 msec of delay jitter, and 3% of packet loss to support an acceptable (i.e., a MOS score of 4.0) quality of voice transmission. For example, with 20 msec of delay budget in each of the call access and delivery LANs, only 110 msec is left as the tolerable delay for the intercountry IP link of the global VPN. It is therefore necessary to actively or passively monitor [3] the intercountry IP links of the global VPN using the IP network monitoring tools and utilities (see, e.g., IETF's RFC 2151) to guarantee the QoS.

In active monitoring, emulated services (e.g., phone calls) between enterprise sites of interest over one or more in-service intercountry IP links must be introduced so that the peak and average values of parameters such as dial-tone delivery and call setup delays, one-way delay, delay jitter, and packet loss can be measured. Since these measurements introduce additional traffic in the IP links and other network elements (such as MGWs, call servers, and routers), it is wise to perform these types of tests over several hours unless it is absolutely necessary to do so at one time.

In passive monitoring, special hardware devices or software probes and processes such as simple network management protocol (SNMP) traps are embedded in the network elements to collect information on packet delay, dispatch rate, loss, and so on. Additional information on routing and transmission of call setup, media, and management of traffic (or packets) in routers, switches, VoIP GWs, call servers, and so on is also collected. These statistics can be retrieved and analyzed periodically from the SNMP management information base (MIB) for network performance monitoring and capacity planning purposes. This type of monitoring is more commonly used in enterprise networks.

It has also been suggested that voice calls be routed over low-hop-count (or fewer node) paths [4] in order to guarantee higher transmission quality. This strategy attempts to minimize the number of nodes in the path from the caller's access LAN to the called party's (i.e., call delivery) LAN, and hence effectively reduces the number of network elements where the packets may suffer queue-ing-related impairments such as delay, delay jitter, and dropping or discarding.

In general, both active and passive monitoring of network performance call for deployment of additional SLA monitoring servers and software tools for processing the information obtained via SNMP probes or traps, periodically executing "ping" and "trace-route" commands to measure the round-trip time, the number of hops needed to reach a destination, and so on. Therefore, additional resources need to be allocated for these hardware and software plat-forms.

The network performance–related information collected using these additional tools is utilized to make intelligent call routing decisions, to guarantee the QoS, and to dynamically update the list of cost-effective alternate or standby intercountry IP links for the global VPN. These additional investments not only allow corporations with global operations to use the same network for real-time multimedia communications, but also help them unify network infra-structures, as well as their operations and managements [5]. Note that the same network can be used for intrasite and intersite wireless communications as well [6] with proper planning [7] and appropriate investments in required infra-structures such as wireless base stations, cordless handsets, and so on [6].

## VoIP FOR CONSUMERS' INTERNATIONAL TELEPHONE CALLS

Implementation of a VoIP-based international telephone calling service for the residential PSTN customer is conceptually similar to the realization of the IP-based long-distance (LD) telephone service within national boundaries, as discussed in Chapter 7. It is possible to use the architecture shown in Figure 7-1 with the following modifications to introduce this service: (a) the Intranet or VPN should be a global Intranet or a global VPN with intercountry IP links, (b) the SS7 signaling gateway (SG) should support the local variants of the SS7 signaling, such as, ASNI-SS7-based signaling in the United States, ITU-T-SS7-based signaling in Europe, country-specific variations of ITU-T-SS7 signaling, and so on, and (c) the IP-PSTN MGWs should support the local variants of channels or links, such as T1 and T3 in the United States, E1 and E3 in Europe, and so on. The modified system architecture is as shown in Figure 8-3.

The VoIP-based international telephone calling service providers can estab-lish one or more operations centers in each country where they wish to sell their telephone calling and other related services. These operations centers are com-monly known as the *point of presence* (POP) in each country. The network elements installed in these POPs are very similar to those used in the network operations centers of multinational corporations—which support IP-PBX- and VoIP-based international calling services—as discussed in the previous section. Additional functionalities or network elements in these POPs may include one or more of the following:

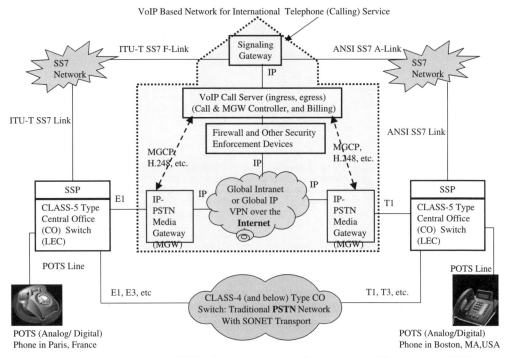

**Figure 8-3** Deployment of VoIP for an international telephone (calling) service (TDM or circuit-switched link, e.g., T1/E1-CAS/PRI, E1/E3-IMT, T1/T3-IMT; IP: IP-based link). (Source: Adapted from Fig. 7-1)

a. Automatic call distributors (ACDs) to resolve billing and other service-related complaints from customers by using the IVR system or by routing the calls to customer service representatives (CSRs);

b. Additional servers to support user authentication, billing, and security services for calling card–based international calling;

c. IP-based advanced applications and feature servers to introduce emerging services efficiently, as shown in Figures 7-4 and 7-6; and

d. Enhanced capabilities of the MGWs and SGs mentioned at the beginning of this section.

Figure 8-4 shows the high-level organization of the network elements within such a POP. For a small-scale operation, the SS7 SG may not be needed initially, as long as the IP-PSTN MGWs support ISDN-PRI- and T1-CAS-type links for PSTN connectivity to a POP in North America, ISDN-PRI- and E1-type links for PSTN connectivity to a POP in Europe, and so on. Note that the call setup performance is usually significantly better when intermachine trunk (IMT) and ISDN-PRI-type links are utilized to support PSTN connectivity.

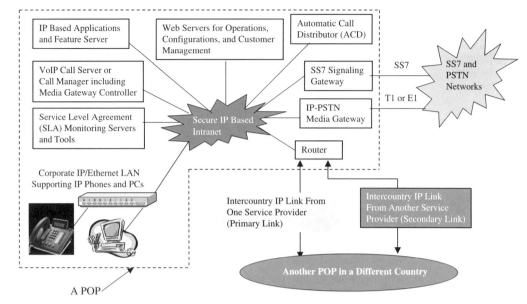

**Figure 8-4** The network elements that are needed in a POP and their interconnection to support an VoIP-based international telephone (calling) service for residential customers.

However, the SS7 SG must be deployed in the POP to use the IMT-type trunks to connect the IP-PSTN MGW to the PSTN.

The VoIP call server or call manager should be dimensioned as per the call setup request processing capacity (e.g., 100 calls/sec) requirements. Since the call manager is the most critical network element within a POP, there should be at least one standby call manager for every in-service (or operating) call manager in a POP. The same mode of operation should be used for authentication, security processing, and billing servers as well. These networking and call processing elements can be centrally located in one POP to serve the customers over a wide geographical area. The optimum location can be determined by solving the classical facility location problems that are commonly discussed in topology and network design handbooks [7].

The IP-PSTN MGWs and all other network elements within a POP can operate in a load-shared mode (e.g., in a clustered environment) over one location or over multiple geographically adjacent locations in order to support reliable media transmission and other call processing services using shared facilitates.

The intercountry IP links should be continuously monitored using the SLA monitoring techniques discussed in the previous section [3], and should be dimensioned to support the number of international VoIP calling minutes sold (over a specific time period) to the customers for telephone calling between any two specific countries. It may also be helpful to maintain at least two—for example, primary and secondary—IP links, as shown in Figure 8-4, between any two specific countries. The primary link should maintain a direct or low-hop-count connection [4] to support a higher quality of voice transmission, and the secondary one could be of lower ETE capacity and could have a varying number of intermediate nodes. These IP links can operate either in load-sharing mode or one as active and the other (e.g., the one with lower capacity) as standby, so that the continuity of the calling service can be maintained even during minor outage of the transmission facilities.

## EPILOGUE

VoIP-based global communications are a reality today for international calling among both employees of multinational corporations and residential PSTN customers. National long-distance carriers and international calling service providers are deploying this service on a limited scale for both multinational corporations and residential customers.

In PSTN networks, the availability of a dial tone is guaranteed within 300 msec of picking up the handset in 95% of the instances, as mentioned in the LSSGRs; call setup delays are at most 3 sec and 10 sec for local and international calls, respectively, after the last digit is entered; and toll quality (i.e., a MOS score of 4.0) of voice transmission is almost always guaranteed. It may be difficult to support cost-effectively the traditional PSTN-grade availability, reliability, and security for calling services to hundreds of thousands of customers using IP-based network elements for call control and signaling and media transmission.

Both competitive and traditional telephone service providers, however, are rolling out VoIP-based national and international calling services using many innovative solutions, including (a) using one-for-one redundancy for VoIP call servers or call managers and other critical network elements in a POP, (b) clustering of IP-PSTN MGWs and other network elements to provide shared protection of services, (c) peering of network nodes and links to maintain an acceptable level of QoS for packet transmission, and (d) active and passive monitoring of network and nodal resources such as transmission and call-processing capabilities so that the toll quality of voice transmission can be guaranteed for the admitted voice connections.

We expect to see further proliferation of these types of networking and service protection techniques for VoIP and related services in the next-generation public and enterprise networks within 10 years.

## REFERENCES

1. W. Stallings, Business Data Communications, Fourth Edition, Prentice-Hall, Upper Saddle River, New Jersey, 2001.

2. G.114 Recommendation, One-Way Transmission Time, ITU-T, Geneva, 1996.

3. T. Chen, Guest Editor, "Network Traffic Measurements and Experiments," Special Feature Topic Issue, IEEE Communications Magazine, Vol. 38, No. 5, pp. 120–167, May 2000.

4. M. Baldi and F. Risso, "Efficiency of Packet Voice with Deterministic Delay," IEEE Communications Magazine, Vol. 38, No. 5, pp. 170–177, May 2000.

5. B. Khasnabish, "Next-Generation Corporate Networks," IEEE IT Pro Magazine, Vol. 2, No. 1, pp. 56–60, January–February 2000.

6. Y.-B. Lin, B. Khasnabish, and I. Chlamtac, "The Wireless Segment of Enterprise Networking," IEEE Network, Vol. 12, No. 4, pp. 50–55, July–August 1998.

7. T. G. Robertazzi, Planning Telecommunication Networks, IEEE Press, New York, 1999.

# 9

# CONCLUSIONS AND CHALLENGES[1]

The technologies and standards required to implement VoIP service are currently well established and mature. Many equipment manufacturers are presently marketing interoperable products—such as IP phones, call servers or managers, MGWs and SGs, and so on—to realize and manage the VoIP service. Advances in digital signal processing and networking software, hardware, and protocol technologies—as discussed in Chapters 2 and 3—have stimulated the development of plug-and-talk-based IP phones. These phones can support VoIP service over IP-based networks equipped with a call controller (CC) and MGW. As discussed in Chapter 6, IP phones can automatically register themselves to a DNS/ENUM server once they are connected to an IP-based network. Furthermore, these phones can derive electric power from the highly reliable Ethernet switches over those wires of the Ethernet cable (a category 5 cable) that are not being used for data services. The software and hardware configurations needed to deliver the subscribed services to the IP phones can be managed—that is, upgraded, added, or deleted—by the subscribers themselves, and the required configurations can be downloaded over the Web by the customers. Because of the openness and flexibility of IP-based call control and signaling, many new and advanced add-on services can be developed, integrated, and marketed very quickly and cost-effectively. VoIP service can be implemented either in a standalone fashion or as a complement to existing PSTN-based telephone calling service.

When VoIP service is implemented in a standalone fashion, the following

---

[1] The ideas and viewpoints presented here belong solely to Bhumip Khasnabish, Massachusetts, USA.

elements are needed: IP phones, various computer servers for DNS/ENUM, call control and signaling, hosting of applications and features, Ethernet switches, IP-based edge and core routers, wiring infrastructures that are commonly used in LANs with an option to deliver electric power to the phones from the switches, and so on.

In order to maintain the interoperability of the standalone VoIP network with the well-established and century-old PSTN endpoints or terminals (i.e., POTS phones), a number of server and GW devices are necessary. These devices include (a) IP-PSTN MGWs to support interworking with PSTN transmission infrastructures (such as TDM links and trunks), (b) an SS7 SG to interact with PSTN signaling and call control infrastructures, (c) service and feature extraction and insertion servers and/or GWs for delivering the caller's name and number, voice message, and so on to an IP phone from a PSTN-hosted feature server and voice mail box, respectively, and (d) GWs and/or servers to support interactions with PSTN billing, operations, and management systems.

Most enterprise networks have IP-based networking infrastructure already in place for intracompany distributed computing and communications. These networks are ideal candidates for experimenting with the introduction of VoIP and other related services. Some medium-sized and large enterprises are already experimenting with or have limited deployment of VoIP service, and they are using many of the methods and architectures discussed in Chapters 6 and 8.

In public networks, VoIP service is now being introduced for national long-distance (LD) and international telephone calling services, as discussed in Chapters 7 and 8. Traditionally, these services are offered by using the existing CLASS-4 (and lower) type PSTN switches and SONET rings. These PSTN switches are currently being replaced or augmented by IP-based multiservice switches and routers that can support both voice and data communications. The SONET rings are being replaced or augmented by IP-based mesh networks of core (gigabit or terabit speed) routers, and these routers can support IP over SONET (packet over SONET) or IP directly over fiber (e.g., gigabit Ethernet for trunking applications) channels [1].

For residential telephone services, CLASS-5 switch replacement may not occur in the foreseeable future. This is due to the fact that existing assets for CLASS-5 switching and the twisted-pair copper wire–based local access network have not yet fully depreciated even in the developed countries. Consequently, service providers are delivering VoIP service to residential customers using DSL lines, CATV networks' channels, wireless local loop (WLL) channels, and so on, as discussed in Chapter 7.

Based on our experiences and experiments (as discussed in Chapters 4 and 5 and the appendixes), we present a set of guidelines for rolling out VoIP services using any operational IP network. This is followed by a discussion of the most challenging future research topics on the implementation of VoIP service.

## GUIDELINES FOR IMPLEMENTING VoIP

As mentioned throughout this book, the openness, flexibility, and widespread availability of the IP-based network and services are fueling the unification of distributed computing and communications (datacom and telecom) infrastructures, and their management and operations [2]. These promote consolidation of the network element procurement and maintenance processes, resulting in a significant reduction in overhead for networks and services.

However, in order to achieve cost-effective (i.e., profitable) implementation of the VoIP service, one must carefully select the following:

a. An open and scalable *architecture framework* for both networking and enhancing the VoIP service;

b. Easily configurable standard interfaces and simple *protocols* for interactions and interconnects among the various networking, service hosting, and user-domain elements. This should be followed by the *design of access and transport networks* and a plan to *upgrade the existing operations support system*; and

c. Mechanisms to guarantee—that is, assign or allocate, monitor, and maintain—user authentication, secure transmission, and ETE *QoS* that are at least as good as those supported by the PSTN.

As discussed in Chapter 1, the multiplane architecture framework proposed by the MSF (as shown in Fig. 1-9 of Chapter 1) can be utilized for separating the call control, media adaptation, and application hosting functions. This architecture also helps achieve the required level of openness (of the interconnections) and granularity (of the elements), that makes it highly scalable. IETF, ISC, and Cable Labs (www.cablelabs.com/projects/) have proposed similar architecture frameworks. Once an architecture framework is selected, a multiphase service rollout plan can be designed for a graceful transition of the existing network infrastructures to deliver telephony and data services over an IP-based, unified communication system.

For IP phones, although H.323, MGCP, and SIP (these protocols are discussed in Chapter 3) based phones are available, the SIP (IETF's RFC 3261) phones are gaining significantly more acceptance in the user community. This can be attributed to the fact that SIP exploits a simple request/response protocol, such as the clear-text based instructions and headers like those in the hypertext transfer protocol (HTTP), and many other Internet-based methods and protocols (DNS, MIME, URL, SMTP, etc.) for setting up a real-time telephone conversation session. These features also enable server-based deployment and low-overhead invocation of many popular add-on services by the SIP phones (clients). These services include Web-based click-to-call, unified messaging, instant messaging and conferencing, on-line transactions (buying/ selling), and so on.

SIP-based VoIP service can be introduced in any operational IP network of an enterprise by deploying the SIP server-based VoIP call controller (CC), call manager, or call server. The SIP server contains an SIP registrar, a proxy, and a redirect server, and the SIP phone contains the user agent—both client and server. In addition to the basic call features (see, e.g., Table 6-2 in Chapter 6), the VoIP CC hosts the basic call processing functions and maintains the call states. Within the enterprise, a call between two SIP phones can now be established under the control of one or more VoIP CCs of the enterprise network, as illustrated in Chapters 3 and 6. In its simplest form, the VoIP CC may be equipped with (a) a VoIP GW (a line card) with T1 interfaces for connectivity with the PSTN network, for example, and (b) a line card to support the SMDI interface over analog lines for integration with the existing voice mail system, for example. Alternatively, (a) a line card–based VoIP GW can be introduced in the existing (circuit-switch-based) PBX chassis with its Ethernet ports supporting connectivity to a corporate LAN, and (b) an adjacent voice mail GW device can be utilized for integration with an existing voice mail system over the SDMI interface. To support unified messaging in such an environment, additional servers—which support open APIs such as TAPI and JTAPI and VoiceXML for IVR scripting—can be introduced. XML-based scripting can be utilized for managing user profiles and the system configuration over the Web [3].

As illustrated in Chapter 7, for residential customers the IP telephony service can be delivered over DSL modem, cable modem (CM), and WLL receiver–based network connections. Service providers need to be equipped with VoIP call servers to perform call control and signaling for the VoIP calls originating from the SIP phones—attached to the DSL modem, CM, or WLL receiver—on customers' premises. If telephone calls from SIP phones are destined for PSTN or POTS phones, network elements such as the SS7 SG and the IP-PSTN media gateway controller (MGC) will be directly involved in setting up the connection for the call. The IP-PSTN MGW will provide all of the required adaptation of the voice signal (or media) from the PSTN or TDM network to the IP network. These are discussed in detail in Chapters 7 and 8 for various network evolution scenarios. As discussed in Chapter 3, MGCP or the H.248/ Megaco protocol can be utilized to control the IP-PSTN MGW from the MGC, and the MGC software can reside either in the VoIP call server or in an adjacent computer server.

In order to support continuous availability of the service from the VoIP call server, it may be necessary to provide battery-based backup of the electric power supply. It may also be necessary to provide one-for-one redundancy of the VoIP call server and the associated MGC. For the MGWs, applications, and feature servers, cluster-based interconnections can be utilized to implement a scalable solution cost-effectively.

The IP network that is interconnecting these clusters, the VoIP CC, and the access and delivery networks (LANs) must be multiconnected and properly engineered so that it meets or exceeds the ETE QoS requirements for the VoIP service. In addition to the reliability and availability requirements, network requirements for delay jitter, packet loss, and network delay limits—as discussed in Chapter 4—must be satisfied to deliver acceptable voice quality over a VoIP session. As discussed in Appendixes A, B, and C, a number of VoIP related experiments have been conducted using the testbed described in Chapter 5. The results reveal that network impairments such as packet loss and delay jitter significantly affect the transmission of both voice and DTMF signals. Network delay—up to a certain limit—seems to have a less severe impact on voice and DTMF transmission. DTMF transmission does not seem to be strongly affected by network delay. However, excessive delay jitter, packet loss, and network delay sometimes cause call establishment attempts to fail repeatedly.

When a call setup request arrives at the VoIP call server, it should be honored only when sufficient ETE resources are available. These resources include bandwidth, buffers, and processing capacity for setting up and maintaining the RTP session for the duration of the VoIP session.

As far as layer-1 (the physical layer of Fig. 2-10 in Chapter 2) is concerned, two or more different physical links can be maintained from the Ethernet switch (to which a SIP phone may be attached), for example, to the VoIP call server. This strategy ensures that the call setup request will reach the VoIP call server even when the primary physical connection fails.

In layer-2, the priority bits (a 3 bit field), as standardized by the IEEE 802.1p group, and the virtual LAN TAG bytes (a 4 byte field), as standardized by the IEEE 802.1Q group, can be utilized to mark the traffic that is carrying voice signal. These fields can define an appropriate class of service for the voice packets and the user priority (virtual LAN tagging) in the media access control (MAC) sublayer of the link layer (see, e.g., Fig. 2-10). In IEEE 802.1p, eight different traffic prioritization levels are defined at the MAC framing sublayer, and it uses a filtering mechanism to retain the multicast traffic within layer-2-switched networks. The IEEE 802.1Q standard defines a format for tagging the frames. Although the virtual LANs are server-port based, the services can be extended to the desktop via tagging on trunk lines in LAN switches and routers. The IEEE 802.1p standard–based setting of the priority bits works very well with the IEEE 802.1Q specifications for virtual LAN tagging. These features can be exploited to achieve prioritized routing of the voice packets in any Enterprise or private IP network. Most currently available LAN switches support implementation of the IEEE 802.1p and IEEE 802.Q standards. Further details on the activities of IEEE 802 work and study groups can be found at their websites (www.ieee802.org/dots.html, 2001).

In the network layer (layer-3 of Fig. 2-10), the type of service (TOS, an 8 bit field, as shown in Figs. 2-3 and 2-6) byte in IPv4 and IPv6 headers can be utilized for setting the DiffServ code point (DSCP) at the edge of the IP network. This enables classification of the packets into a small number of aggregated flows or classes. Consequently, at each DiffServ router, the VoIP-based telephone conversation related packets could be routed by using the expedited forwarding (EF, RFC 3246 and RFC 3247) technique. This ensures that the traffic from real-time voice conversation will be forwarded without excessive delay, and delay variations have the objective of maintaining acceptable quality (e.g., a MOS score of 4.0) of real-time voice communication. For real-time voice services over a long-haul network, the core or backbone IP network should be capable of supporting multiple MPLS tunnels (as mentioned in Chapter 2) from one edge router (e.g., the access network's) to another edge router (e.g., the delivery network's). Although these MPLS tunnels are implemented over multiples routers, they offer virtual one-hop paths from one edge router to another, which helps maintain a high quality of traffic transmission. In addition, in the core network, the IP packets can be transmitted directly over SONET frames or optical channels, which can support additional robustness of traffic transmission [1].

In the layers above layer-3, the hardware devices and software processes must work in unison in order to support the availability of a dial tone, an acceptable level of call setup delay, and a high quality of voice packet transmission. To achieve this, the IP phone and the CPE or IAD must have the ability to identify the real-time voice session–related packets so that these packets can be emitted with the highest possible priority using a smaller separate hardware-based queue.

In order to set up and maintain a QoS-guaranteed ETE path over an IP network, many of the IETF recommended protocols could be exploited. For example, Cable Labs has defined RSVP- and COPS-based mechanisms (PKT-TR-ARCH1.2-V01-001229, PKT-SP-DQOS-I03-020116, and pkt-sp-iqos-i01-001128 at www.packetcable.com/specifications) to dynamically maintain QoS over a packet cable network for VoIP service. Additional server-based mechanisms can be introduced to authenticate VoIP session users and to encrypt the voice signal in order to ensure secure communications, as described in the packet-cable specifications (see, e.g., pkt-tr-arch-v01-991201 and PKT-SP-SEC-I05-020116 at www.packetcable.com/specifications).

Many other network scalability, service availability, and general interoperability–related problems still exist; these are discussed in the next section. A host of standardization forums are working to resolve these issues, as mentioned in the last section of this chapter.

## VoIP IMPLEMENTATION CHALLENGES

The openness and flexibility of the IP and the World Wide Web (WWW) have steered the development of a variety of devices and techniques for implementation of VoIP. However, the same openness, ubiquity, and flexibility to support open APIs and to carry multiple types of traffic (e.g., real- and non-real-time audio, video, and data streams) have also created many service reliability and security–related concerns. Some of these issues are discussed below, along with a few desirable features of the feasible solutions.

## Simplicity and Ease of Use

The IP phone and the VoIP service should be at least as easy to deploy and use as the traditional (rotary or with a numeric keypad) PSTN phones, irrespective of whether they are implemented over a corporate LAN or to residential customers over DSL, CATV, WLL, or other facilities. The IP phones should be self-configuring, and should be able to automatically download debugging and service upgrade–related software from a designated primary (or secondary) server after appropriate device authentication. Users should be able to invoke the emerging services from an IP phone via a request/response method or through an IVR system. An IP phone should be able to ring another IP phone by dialing either a phone number or an e-mail address.

## Nonstop Service

The VoIP call server should be able to deliver a dial tone to the IP phone and process call setup requests even when there is a failure of the electric power supply at both customers' premises and service providers' buildings. The call server itself should not only have duplicates of all the software and hardware components, it probably should be operating with one hot standby unit or in one-for-one redundancy mode. In addition, it should be able to route an incoming call—that is, establish the bearer path—to the destination as long as the called IP or POTS phone is operational so that the caller can hear the ring-back or busy tone.

## High-Quality Service

The network elements, access and transport IP networks, protocols, and system architecture must work together to deliver high-quality VoIP service. The service should be at least as good as the TDM or circuit switch technology–based voice telephony service traditionally offered by the POTS network (or PSTN). As mentioned in Chapter 4, for real-time telephony service to residential customers, the service starts the moment the handset is picked up by a customer and ends long after the call is completed. The network elements, and the access and transport IP networks, must be designed and configured in such a way that (a) the availability of the dial tone to the phones (IP or other types of phones) can be guaranteed all the time, (b) the quality of voice signal transmission remains high (e.g., a MOS value of 4.0), and (c) appropriate billing and satisfactory customer service can be assured. Sufficiently open network architectures with precisely defined separation of functions should help achieve all of these goals concurrently.

## Scalable Solutions

The combination of network elements, access and transport IP networks, protocols, and system architecture must scale well to support hundreds of thousands of customers without affecting service availability or the quality of transmission (in real time) of the voice signal. Once again, the network architecture and the protocol sets should be carefully selected so that the scalability of both the network and the service is built into the implementations.

### Interoperability

This is necessary to protect the capital that has already been invested in the existing networking and telephony service delivery infrastructures. These infrastructures include both legacy PSTN facilities and the existing (if any) VoIP service delivery systems from different equipment manufacturers. The PSTN facilities are the POTS phones, the PSTN switches, access and transmission TDM networks, and the associated billing, operations, and call feature hosts. In the same way, there may also exist a variety of first-generation IP telephony–related equipment—such as H.323 and MGCP phones and the related GWs, GKs, and billing systems—in both enterprise and public telephone service providers' networks. Therefore, the network elements for implementing or enhancing the existing VoIP service should be selected to support both legacy systems and earlier generations (and versions) of the line (and/or trunk) interfaces and VoIP protocols. These help the network designers to choose the proper software and hardware configurations of the network elements for rapid deployment of VoIP service via incremental evolution of the network.

### Authentication and Security

Because of the openness of IP and the ubiquity of the Internet, IP-based telephony endpoints and VoIP network elements are susceptible to both malicious attacks and inadvertent damage (during configuration change, software upgrade, etc.). To minimize the chances of these events, multiple levels of user (or client) authentication can be utilized before allowing access to or approving a VoIP call over the network. Also, a variety of security enforcement devices such as firewalls, proxy servers, and so on can be utilized to safeguard users' access to the critical networking and service hosting facilities. To achieve secure communication over the shared IP network, large (e.g., 1024 bit) key-based encryption can be utilized. However, this may add further overhead such as maintenance of key distribution centers and clients' need to communicate with these centers for each session, and may cause degradation of the quality of voice signal transmission. Other solutions include utilization of VoIP session setup (e.g., the latest version of SIP; RFC 3261, RFC 3262) and IP communication (e.g., IPv6, RFC 2460) protocols that have built-in security-related services.

### Legal and Public Safety–Related Services

To comply with regulatory requirements, the public telecom service providers who offer VoIP or IP telephony-based basic telephony services to residential customers must implement services like CALEA, routing of 911 calls to the PSAPs with callers' identification and location information, and so on. However, the enterprises may need to implement these services as well once they become fully dependant on IP-based telecommunication services. Since the IP networks operate using shared resources in a distributed fashion, it is sometimes difficult to detect the identities and locations of the communicating parties in a VoIP session. However, many distributed server, wiring diagram, and media access control (MAC address) level filtering mechanisms are currently being explored to resolve these issues.

**Cost-Effective Implementation**

Delivery of voice and data services over multiservice IP networks will not be widespread until and unless the implementation of highly available and reliable VoIP service becomes economically feasible. The network elements that are required to implement the basic IP telephony and VoIP-related services are significantly less expensive (possibly one-tenth) than the traditional PSTN switches. The costs for transmission of a packetized voice signal over shared IP links are also much smaller (e.g., could range from one-tenth to one-fifth) than those of voice transmission over traditional TDM (circuit-switch) networks. This is equally true for both interoffice telephone calls within an enterprise and domestic LD and international telephone calls in public telecom networks. However, the prices of IP phones and the related CPE/IAD, as well as the expenses for implementing PSTN-grade VoIP service in the access networks (e.g., 99.999% of availability of service or 5.256 min of downtime of service per year), may be higher. These higher costs can be attributed, to some extent, to the lack of embedded reliability, security, and QoS features in the software and hardware components and to the protocols that are utilized for implementing the VoIP service. Many work groups are currently developing mechanisms to overcome these limitations.

**EPILOGUE**

Implementation of VoIP promises low-cost realization of telephony and many other feature-rich real-time and non-real-time communications services. However, various issues are posing realistic challenges to network element developers, network designers, and service providers alike. These concerns include managing the PSTN-grade availability, reliability, security, service-completion standards, and customer satisfaction of the telephony service in open and flexible (i.e., IP) networking environments.

Many industry consortiums, task forces, and standardization forums are working with equipment manufacturers and service providers to develop protocols, architecture frameworks, and interoperability specifications to resolve these issues. These include various study groups of ANSI (www.ansi.org) and

ITU-T, work groups of IETF, MSF, and the International Softswitch Consortium (ISC), IEEE (standards.ieee.org), Cable Labs (www.cablelabs.com/projects/), and the DSL and ATM forums. We expect to see publication of practical implementation and interoperability recommendations from these industrywide joint efforts in the foreseeable future.

## REFERENCES

1. B. Khasnabish, "Optical Networking Issues and Opportunities: Service Providers' Perspectives," Optical Networks Magazine, Vol. 3, No. 1, pp. 53–58, January/February 2002.
2. G. Jakobson and B. Khasnabish, Guest Editors, "Enterprise Network and Service Management," Special Issue of IEEE Network Magazine, Vol. 16, No. 1, pp. 6–7, January/February 2002.
3. B. Khasnabish, "Next-Generation Corporate Networks," IEEE IT Pro Magazine, Vol. 2, No. 1, pp. 56–60, January/February 2000.

# APPENDIX A

# CALL PROGRESS TIME MEASUREMENT IN IP TELEPHONY[1]

In IP telephony, a voice call is usually established through multiple stages. In the first stage, a phone number is dialed to reach a near-end or call-originating or ingress IP-telephony (or IP-PSTN) GW. The next stages involve user identification by delivering an $m$-digit user ID to the authentication and/or billing server, followed by user authentication using an $n$-digit personal identification number (PIN). After that, the caller is allowed (a last-stage dial tone is provided) to dial the destination phone number, provided that authentication is successful. This appendix presents a method for measuring call progress time in IP telephony. The proposed technique can be used to measure the system response time at every stage. It is flexible, so that it can be easily modified to include a newly defined tone or set of tones, or a voice-band speech sample can be used at every stage to detect the system's response. The proposed method has been implemented using scripts written in Hammer visual basic (HVB) language (www.hammer.com) for testing with a few commercially available IP-PSTN GWs.

## INTRODUCTION

The first generation of IP-PSTN GWs allows voice call setup in single or multiple stages. The more stages involved, the longer the call setup time. The configuration of a simple testbed, which can be used to measure call setup time, is shown in Figure A-1.

[1] The ideas and viewpoints presented here belong solely to Bhumip Khasnabish, Massachusetts, USA.

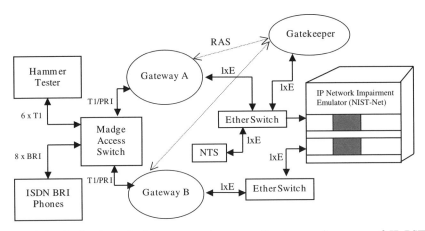

**Figure A-1**  A simple testbed for measuring the call setup performance of IP-PSTN GWs (E: Ethernet link; T1: CAS or PRI link; BRI: basic rate interface links. NTS: network time server; it provides timing information (clock) to IP domain network elements such as IP-PSTN GWs, GK, and NIST-Net, and if needed, it can derive clocking information from a GPS receiver as well. A detailed description of the testbed is presented in Chapter 5.

The Hammer VoIP test equipment [1,2] is used for generating bulk telephony calls and for analyzing the emulated black (or PSTN) phone to black phone calls. This includes measuring the answer time, the response time at various stages of call progress, and the time needed to hear the ring-back tone at the call-originating side. The version of the Hammer tester we used in this testbed can support a maximum of six T1 lines to a PSTN switch or PBX (e.g., a Madge Access Switch).

The ISDN BRI phones can be used to check the integrity of call progress and human perception–based audio quality measurement. Call progress integrity evaluation involves hearing the generation of appropriate tones (e.g., a string of DMTF digits, a dial tone, a ring tone, etc.), playout of an appropriate interactive voice response (IVR) message, and so on. The Madge (www.madge. com, 2001) Access Switch is a small PBX and emulates a CLASS-6-type PSTN central office (CO) switch. It provides one or more T1-CAS and/or T1-PRI connection(s) to the PSTN-side interface(s) of the IP-PSTN GWs under test. A set of ISDN BRI phones can be also directly connected to it.

The two 24-port EtherSwitches and the IP network impairment emulator, which is a PC-based simple router (described in Chapter 5), comprise the Intranet of the testbed. The EtherSwitches provide connectivity to the IP-side interface(s) of the IP-PSTN GWs under test.

GW-A and GW-B are the near-end (call-originating or ingress) and far-end (call-terminating or egress) GWs. Usually they are connected to two different subnets, which are interconnected via the simple PC-based router mentioned above. However, when necessary, it is also possible to connect the two GWs using the same subnet, that is, to connect them to the same EtherSwitch. The gatekeeper (GK) usually runs on a WindowsNT server or a general-purpose IP router and is connected to the same subnet to which the GW-A is connected. In general, the GK performs registration, authentication, and status (RAS) monitoring functions when a call establishment request arrives. If implemented, it can also maintain the call detail record (CDR) files. The network time server (NTS) provides timing information (clock) to the IP domain network elements such as IP-PSTN GWs, GK, and NIST-Net and, if needed, it can derive clocking information from a global positioning system (GPS) receiver as well.

The Madge Access Switch used in the testbed can accommodate a maximum of six 4- or 8-port line cards, with 4 ports in one card reserved for local/remote configuration, network management, and timing management. The remaining ports can be used for BRI and/or T1 (CAS or PRI) connections. Currently, we are using two 8-port cards for connections to BRI phones and the other ports to support T1 connections. The six T1 ports of the Madge Access Switch are connected to the AG-T1 cards of the Hammer using T1-CAS lines. The remaining T1 ports of Madge can be used to connect to either one or three pairs of the GWs under test. Appropriate dialing plans and Madge configurations are used to make telephony connections from one Hammer channel or BRI phone to the other, either directly through Madge or using one or two IP-PSTN GWs. These options offer the flexibility to make calls over the PSTN network/switch alone or through the IP network with incorporation of very little (i.e., when both IP-PSTN GWs are connected to the same subnet) or a controlled amount of impairments like delay jitter, packet loss, bandwidth restrictions, and so on. These impairments are added using an IP network impairment emulator, NIST-Net (described in Chapter 5).

The call processing performance of an IP-PSTN GW can be described in terms of the following two factors. The *first* one is the total amount of time it takes to set up a call, measured from the moment the last digit of the first-stage dialing number is entered to the time the ring-back tone is heard at the call-originating side. This is known as the *call setup time*, and it is discussed in this appendix. Various components of the call setup time are shown in Figures A-2 and A-3. The *second* factor is the number of simultaneous calls that can be handled by a GW without any precall wait. Note that the precall waiting time can vary from as little as 1 sec to as much as 10 sec. This is discussed in Appendix B.

## DESCRIPTION OF THE TECHNIQUE

The technique discussed here consists of two phases of evaluation. The *first phase* involves determining the tone(s) indicating the completion of one stage of dialing so that the beginning of the next stage of dialing can be detected. If an IVR file is played then, any indication of "voice begin" can be used as an indication of the beginning/end of a stage. For example, in some implemen-

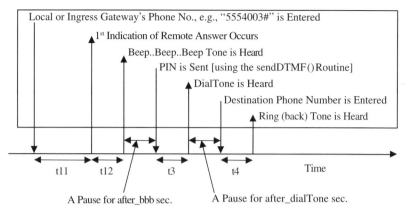

**Figure A-2**  Multistage call setup using PIN-based caller authentication. The user identification stage (response time, t2) is not shown here. The call setup time is computed as t11 + t12 + t3 + t4. The after_bbb and after_dialTone pauses allow stabilization of the system's response.

tations, one or more DTMF digits (e.g., a series of 7s or 9s) are used to indicate the first stage's response, while in others, DTMF digit(s) followed by a continuous tone—which could be a modified or bona-fide dial tone—are used. It is therefore important to determine this tone, either from the manufacturer's specifications or through trials and testing. We utilize some rudimentary addition, detection, and tuning of the tones to detect the readiness of the next stage (of dialing) to accept the incoming digit strings. The upper and lower frequencies and the tolerances (which can vary from ±1 to ±100 Hz) at the boundary

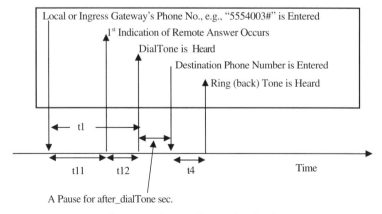

**Figure A-3**  Two-stage call setup where caller authentication is not needed. The call setup time is computed as t11 + t12 + t4. The after_dialTone pause allows stabilization of the system's response.

of a DTMF tone, and its detection widow size (which can vary from 2 to 10 msec or more) need to be properly adjusted to use this feature effectively. Note that in the United States, the *ring tone* is defined as a combination of two tones (frequency 1 = 440 Hz, frequency 2 = 480 Hz) with on-time of 2 sec and off-time of 4 sec, the *dial tone* is a combination of two tones (frequency 1 = 350 Hz, frequency 2 = 440 Hz) played continuously, and the *busy tone* is a combination of two tones (frequency 1 = 480 Hz, frequency 2 = 620 Hz) with on-time of 0.5 sec and off-time of 0.5 sec.

In the *second phase*, it is necessary to read the exact timestamps (the finer the resolution, the greater the accuracy of measurement) of the required telephony events. These timestamps are used to measure the elapsed time between various call progress events. For example, the time difference between the telephony event "last digit entered" in the first stage of dialing and the telephony event "first indication of remote answer" determines the response time of the first stage.

The "first indication of remote answer" usually follows one or more tones or playout of a voice prompt. This indicates that the system is now ready to accept the PIN and/or user ID from the caller to authenticate the caller. This information can also be used for billing or/and call routing purpose(s).

Once the authentication stage is complete, the caller is prompted with a second dial tone, and that's when the caller enters the telephone number (4-digit, 7-digit, or 10-digit, as required) of the destination phone, that is, the called party's terminal. If a valid destination phone number is dialed, the calling party can expect to hear the standard ring-back tone. However, since the transmission medium is IP, a correct/precise ring-back tone may not be heard at the call-originating side. Once again, it is important to find out the exact definition of the ring-back tone either from the manufacturer's specifications or through trials and testing. Note that all of the standard test and measurement equipment is calibrated or tuned to detect the standard ring-back tone, but none of the IP-PSTN GWs may be capable of delivering it unless a very-high-quality digital phone (e.g., an ISDN BRI phone) is used by the called party.

A description of time measurement in each of the stages is now presented, along with a definition of each telephony event. We are assuming that each event has an embedded member (as available in HVB) for extracting the timestamp with an acceptable level (e.g., milliseconds or less) of resolution.

In the first stage, a 7- or 10-digit telephone number is entered to *reach the call-originating* (ingress or near-end) IP-PSTN *GW*. The response time for this stage is the difference between the time the last digit of the digit string is entered and an indication of an answer from the remote end (i.e., the IP-PSTN GW). Note that if the "indication of remote answer" is the playout of an IVR message, it is necessary to wait until the voice playout is completed. This sequence is shown in Figure A-4.

Next, it is necessary to determine the response time needed to allow the user to provide input for *identification* via a user-id (4 to 8 digits, for example) and then get a result. The response could be a standard dial tone, a vendor-specific

1   Set telephonyEvent = placingACall (a number with # sign at the end, e.g.,
    97814662080#)
2   Set telephonyEvent1 = getCallEvent (ALL_DIGITS_SENT)
3   Set telephonyEvent2 = getCallEvent(REMOTE_ANSWERED)
4   Set telephonyEvent3 = getCallEvent(TONE or Voice_Begin)
5   If voice_begin is used, one must wait for voice_end before beginning the next
    action, and for detecting the remote_answer_tone, tolerance of DTMFs and/
    or the detection window size may need to be adjusted
6   FirstStageResponseTime, t1 or (t11 + t12) = (telephonyEvent3.time –
    telephonyEvent1.time)

**Figure A-4**   Description of the measurement of first-stage response time.

tone or set of tones, or simply the beginning of a voice prompt announcing that
the caller is now in the identification phase. Here again, the identification of
"voice begin" may be easier than pinning down the tone or set of tones, especially when IP transport is used. This sequence is shown in Figure A-5.

If the identification is successful, the caller can proceed to the next stage, and
an IVR or dial tone will be heard. A failed identification may result in a busy
tone or a different IVR message.

It is then necessary to determine the response time needed to allow the user
to provide input for *authentication* (e.g., via the use of a 4- to 8-digit PIN). The
response could be a standard dial tone, a vendor-specific tone or set of tones, or
simply the beginning of a voice prompt announcing the result of authentication. Here again, the identification of "voice begin" may be easier than pinning
down the tone or set of tones, especially when IP transport is used. The steps
are shown in Figure A-6.

If the authentication is successful, the caller can proceed to the next stage,
and an IVR or dial tone will be heard. A failed authentication may result in a
busy or fast-busy tone or a voice announcement.

1   Set telephonyEvent4 = sendingADtmfString (a string of digits with # sign at
    the end, e.g., "1234#")
2   Set telephonyEvent5 = getCallEvent(REMOTE_ANSWER_TONE or
    Voice_Begin)
3   If voice_begin is used, one must wait for voice_end before beginning the next
    action, and for detecting the remote_answer_tone tolerance of DTMF and/or
    detection window size may need to be adjusted
4   SecondStageResponseTime, t2 = (telephonyEvent5.time – telephonyEvent4.
    time)

**Figure A-5**   Description of the measurement of second-stage response time.

1   Set telephonyEvent6 = sendingADtmfString (a string of digits with # sign at the end, e.g., "5678#")
2   Set telephonyEvent7 = getCallEvent(REMOTE_ANSWER_TONE or Voice_Begin)
3   If voice_begin is used, one must wait for voice_end before beginning the next action, and for detecting the remote_answer_tone, tolerance of DTMF and/or detection window size may need to be adjusted
4   ThirdStageResponseTime, t3 = (telephonyEvent7.time − telephonyEvent6.time)

**Figure A-6**   Description of the measurement of third-stage response time.

Finally, the caller hears the dial tone or an IVR message so that now the *destination number* (i.e., the called party's E.164 address) can be *dialed*. After the destination number is dialed, a ring-back tone or busy tone should be heard. The time interval between entering the last digit of the destination number and hearing the ring-back or busy tone is the response time for this stage of call establishment. This sequence is shown in Figure A-7.

At this point, the nonbusy called party answers the telephone call (i.e., picks up the handset) and the connection is established. Once the connection is up, the calling party hears the called party's voice, which is delayed by the one-way voice transport delay (voice envelop delay or speech latency). This delay has a significant impact on the voice quality. The smaller the delay, the better the voice quality.

**An Implementation Using HVB Language**

This section presents an implementation of the proposed technique using HVB [1] language. It consists of the following two routines: placeCall.sbl and rcvCall.sbl. The placeCall.sbl routine emulates a calling party. It implements

1   Set telephonyEvent8 = sendingADtmfString (a string of digits for destination telephone number with # sign at the end, e.g., "17814662130#")
2   Set telephonyEvent9 = getCallEvent(GW-Vendor_Specific_RING_TONE or BUSY_TONE)
3   For detecting the ring_tone or busy_tone, tolerance of DTMF and/or detection window size may need to be adjusted
4   FourthStageResponseTime, t4 = (telephonyEvent9.time − telephonyEvent8.time)

**Figure A-7**   Description of the measurement of fourth-stage response time.

```
'***---- FileName: placeCall.sbl → This routine is used to place calls through an
IP-PSTN GW

waitTick = 1                    '***---- set waitTick to zero if all calls are to be
                                started simultaneously
sendID = chanId( ) + 24    '***---- to determine the Id of the called channel

'***---- Use the following to dial from Hammer channel to Hammer channel
dialnumber = "55720" + str(sendID) + "#"
'***---- Use the following to dial from Hammer channel to ISDN-BRI phone
'dialnumber = "7815573004#"

call startProtocol (HT_PROTO_WNK0,,)
waitTime = (sendID-24)* waitTick
pause waitTime, HT_SECONDS
logmsg "⇒ Pre-Call Wait is " & waitTime & " sec", HT_LOG_DEBUG

'***---- Placing a call using the telephone number of the ingress GW
  logMsg "⇒ Placing call to 5554001", HT_LOG_DEBUG
  set event = placeCall ("5554001#",)
  set event1 = getCallEvent(HTCALL_DIGITS_SENT)
  set event2 = getCallEvent(HTCALL_REMOTE_ANSWERED)
```

*resTime = event2 − event1*   '***---- *Computation of t11, the first stage response*
*time*
```
logmsg "⇒ First stage response time (t11) is " & resTime & " msec", HT_LOG_
DEBUG
  eventType = event.type( )
  if eventType <> HTEVT_CALL_CONNECTED then goto TelError

'***---- Waiting for the beginning or start of IVR message
set event3 = WaitForIvrStart( )
```
*resTime = event3 − event2*   '***---- *Computation of t12, the first stage (2nd half)*
*response time*
```
logmsg "⇒ First stage (2nd half) response time (t12) is " & resTime & " msec",
HT_LOG_DEBUG

'***---- Waiting for the end of IVR message
set event = WaitForIvrEnd( )
  eventType = event.type( )
  if (event.value( ) <> HT_CP_VOICE_END) then
    logmsg "⇒ Error in IVR playout ! "
    goto TelError '* can set CallFailedReson stats here
  end if
```

**Figure A-8a** A segment of the placeCall.sbl routine written in HVB language. This script emulates a calling party over an analog line (using wink_start0 protocol). It dials a local or ingress IP-PSTN GW and then detects the voice signal before proceeding to the next stage of dialing for call setup. At every stage it measures the response time by subtracting the time of occurrence of the telephony events.

```
'***---- The second stage is not needed for type-L GW, and hence it is not shown
here.
***---- Note: The second stage response time is t2 = (event5 − event4)
'***---- Entering the PIN number for caller authentication
   logMsg "⇒ Sending DTMF digits 42#", HT_LOG_DEBUG
   set event6 = sendDtmf("42#",)
   eventType = event6.type( )
   if eventType <> HTEVT_TONES_DONE then goto TelError

'***---- Waiting for the beginning or start of IVR message
set event7 = WaitForIvrStart( )
resTime = event7 − event6   '***---- Computation of t3, the third stage response
                               time
logmsg "⇒ Second stage response time (t3) is " & resTime & " msec",
HT_LOG_DEBUG
```

**Figure A-8a** *(Continued)*

various telephony events needed to make a call setup request. The required telephony events consist of placing a call and waiting for a response. Next, it is necessary to send a set of digits for user identification and then wait for a response. Entering a set of digits for user authentication and waiting for a response follows. The final step is to enter the destination phone number and wait for the ring-back tone occur. If the connection request is successful, a set of voice prompts must be played, followed by release of the call. Figures A-8a, A-8b, and A-9 present HVB-based implementation of the above function-alities. Figures A-10a and A-10b show implementation of various functions for detecting the call progress events and the beginning and end of voice prompt playout. The rcvCall.sbl routine emulates a called party. It implements tele-phony events such as answering a call after a prespecified set of rings, playing one or more voice prompts, and then releasing the call. The implementation is shown in Figure A-11.

For example, the placeCall.sbl script can be executed on the first channel of the first AG-T1 board (Board no. 0) of the Hammer tester (see Fig. A-1), and the rcvCall.sbl can be executed on the first channel of the second AG-T1 board (Board no. 1) of the same Hammer tester.

## RESULTS

As mentioned earlier, we have implemented the proposed technique using the HVB language, and it is fully functional in Hammer's integrated telephony (HammerIT) tester running the HammerIT 2.1.3 operating system.

```
'***---- Waiting for the end of IVR message
set event = WaitForIvrEnd( )
   eventType = event.type( )
   if (event.value( ) <> HT_CP_VOICE_END) then
      logmsg "⇒ Error in IVR playout ! "
      goto TelError '* can set CallFailedReson stats here
   end if

'***---- Entering the destination telephone number
   logMsg "⇒ Sending DTMF digits" & dialnumber, HT_LOG_DEBUG
   set event8 = sendDtmf(dialnumber,)
   eventType = event8.type( )
   if eventType <> HTEVT_TONES_DONE then goto TelError

'***---- Detect Ringtone, when calling to the BRI phone
'***---- use 3 msec detection window when calling BRI phone
'***---- use 1 msec detection window for Hammer-to-Hammer Calls

toneDetParam.setVal HTPARM_TDET_TIME, 1
' addTone 3,1500,100,0,0, toneDetParam    '* tone from Hammer
addTone 3,440,5,480,5, toneDetParam       '* ringtone from the BRI phone
   set event9 = WaitForCPEvent( HTEVT_TONE_3_BEGIN )
   eventType = event9.type( )
   if eventType <> HTEVT_TONE_3_BEGIN then
      logMsg "!! No ringtone after the 4th-stage of dialing"
      goto TelError '* can set CallFailedReson stats here
   end if
   removeTone(3)
```

*resTime = event9 − event8   '***---- Computation of t4, the fourth stage response time*

```
logmsg "⇒ Time to hear RingTone (t4) after IVR-Finished " & resTime &
" msec", HT_LOG_DEBUG

pause 10, HT_SECONDS   '***---- allow a 10 sec talk time from the called
party

'***---- Set the value of repeatPrompt to 100 to emulate a ~10-minute
conversation
For repeatPrompt = 1 to 200 step 1
   clearDigits
   logMsg "⇒ Playing voipwom1p4.pcm", HT_LOG_DEBUG
   set event = playPrompt("voipwom1p4.pcm", HT_ENCODE_PCM8M16,
   10000,)
   eventType = event.type( )
```

**Figure A-8b**   A segment of the placeCall.sbl routine, which emulates the fourth stage of dialing. It also measures the response time by subtracting the time of occurrence of the telephony events. It then emulates playing of a set of voice prompts with pauses to accommodate play-out of the called party's voice prompts. This is repeated for a pre-specified number of times, depending on the duration of the call.

```
    logMsg "⇒ PlayPrompt has been repeated=" & repeatPrompt & "=times",
    HT_LOG_DEBUG
    if eventType <> HTEVT_PLAY_DONE then goto TelError
    pause 10, HT_SECONDS   '***---- allow a 10 sec talk time from the called
                                  party
Next repeatPrompt

    pause 10, HT_SECONDS
```

**Figure A-8b**  *(Continued)*

We used the above-mentioned technique to measure the call setup time using a variety of IP telephony GWs. In our lab, a Madge Access Switch is used to emulate the PSTN, and EtherSwitches—connected through a NIST-Net router—are used to emulate the Internet. ISDN PRI link(s) are used to connect the GWs to the PSTN in order to support the calls from the ISDN (BRI)

```
set event = releaseCall( )
    eventType = event.type( )
    if eventType <> HTEVT_CALL_RELEASED then goto TelError
    logMsg "⇒ Call released.", HT_LOG_DEBUG
    logMsg "⇒ Call released."

    setScriptResult HT_SUCCESS
goto done

TelError:
    setScriptResult HT_FAILURE
'***---- Add other options here
Exit sub

PrintError:
    logMsg "⇒ Script Failure at line: " + str$(Erl)
    logMsg "⇒ with error: " + Error$
'***---- Add other options here
setScriptResult HT_FAILURE
    Reset
    Exit sub

done:
logMsg "Script (plcCall.sbl) is now finished running."
    Reset
End sub
```

**Figure A-9**   This segment of the placeCall.sbl routine emulates the release of a call, displays that message and then completes the execution of the script.

```
'***---- Wait for call progress event
Function WaitForCPEvent(cpevent as double) as telEvent
    dim event    as telEvent
    dim done     as integer
    dim cpParams as parmCallProg
    dim eventType as double

    done = 0
    while done = 0
        set event = getNextEvent(50000)
        eventType = event.type( )
        event.eventText HT_eventStr
            logMsg "⇒ type = " & str$(eventType), HT_LOG_NORMAL
            if (eventType = cpevent) then done = 1
    wend

    set WaitForCPEvent = event
End Function
```

**Figure A-10a**    This segment of the placeCall.sbl routine emulates the wait for call progress event. The objective is to exit from the loop when a prespecified call progress event occurs.

phone or from emulated analog phones in the Hammer'IT tester. Ethernet links (10/100 BT) are used to connect the IP-PSTN GWs to the EtherSwitches. We have used T1-CAS links from the Hammer tester to the Madge Access Switch. It is also possible to make calls from the BRI phone to the "call receiving script" running on a channel connected to the Hammer tester.

Table A-1 presents the results of call setup time measurement using one type of commercially available IP-PSTN GWs in an idle system. Note that in an idle system no background calls are in progress; the only call in progress is the one for which the call setup time is being measured. It is possible to run background calls or connection setup processes while measuring the call setup time for a connection. These measurements provide performance results for a busy system.

## CONCLUSIONS

A very flexible method for measuring the call progress time in IP telephony has been presented. In IP telephony, a call usually includes several stages. In the first stage, a phone number is dialed to reach a near-end (ingress call-originating) IP telephony GW. The next stages involves user identification and authentication through delivery of an $m$-digit user ID and then an $n$-digit PIN to the authentication and/or billing server. After that, the caller is allowed

```
'***---- Fuction to detect the Start of IVR Playout
Function WaitForIvrStart( ) as telEvent

dim    voiceBegin   as integer
dim    event        as telEvent

startCallProgress
  voiceBegin = 0
  do
    set event = getNextEvent(20000)
    eventType = event.type( )
    if (event.value( ) = HT_CP_VOICE_BEGIN) then
      voiceBegin = 1
      logmsg "⇒ <<<IVR Play Begins>>>"
    elseif (eventType = HTEVT_CP_DONE) then
      '* callprogress analysis stopped before voice was heard.
      logmsg "⇒ CallProgress Timeout"
      exit function
    end if
  loop while voiceBegin = 0

stopCallProgress
set WaitForIvrStart = event
End Function
'****************************************************************
'***---- Function to detect End of IVR Playout
Function WaitForIvrEnd( ) as telEvent

dim    voiceEnd   as integer
dim    event      as telEvent

startCallProgress
  voiceEnd = 0
  do
    set event = getNextEvent(20000)
    eventType = event.type( )
    if (event.value( ) = HT_CP_VOICE_END) then
      voiceEnd = 1
      logmsg "⇒ <<<IVR Play Ends>>>"
    elseif (eventType = HTEVT_CP_DONE) then
      '* callprogress analysis stopped before voice was heard.
      logmsg "⇒ CallProgress Timeout"
      exit function
    end if
  loop while voiceEnd = 0

stopCallProgress
set WaitForIvrEnd = event
End Function
```

**Figure A-10b**  This segment of the placeCall.sbl routine emulates the wait for voice begin/end. The objective is to exit from the loop when play-out of the voice prompt begins or ends.

**149**

```
'***---- FileName: rcvCall.sbl
'***---- Description: This routine is used to receive calls through type-L
IP-PSTN Gateways.
call startProtocol (HT_PROTO_WNK0,,)

    logMsg "⇒ Waiting for incoming call ...", HT_LOG_DEBUG
    set event = waitForCall(-1)
    eventType = event.type( )
    if eventType <> HTEVT_INCOMING_CALL then goto TelError

'***---- Answering the phone call on the first ring
    logMsg "⇒ Receiving incoming call.", HT_LOG_DEBUG
    set event = answerCall(0)
    eventType = event.type( )
    if eventType <> HTEVT_CALL_CONNECTED then goto TelError

'***---- Set the value of repeatPrompt to 100 to emulate a ~10-minute
conversation
For repeatPrompt = 1 to 200 step 1
clearDigits
    logMsg "⇒ Playing voipman2p2.pcm", HT_LOG_DEBUG
    set event = playPrompt("voipman2p2.pcm", HT_ENCODE_PCM8M16,
    10000,)
eventType = event.type( )
    if eventType <> HTEVT_PLAY_DONE then goto TelError
Next repeatPrompt
'pause 500, HT_MILLISECONDS

    set event = releaseCall( )
    eventType = event.type( )
    if eventType <> HTEVT_CALL_RELEASED then goto TelError
    logMsg "⇒ Call released.", HT_LOG_DEBUG
    logMsg "⇒ Call released."

    setScriptResult HT_SUCCESS
goto done

TelError:
    setScriptResult HT_FAILURE
'***---- Add other options here
Exit sub

PrintError:
    logMsg "⇒ Script Failure at line: " + str$(Erl)
    logMsg "⇒ with error: " + Error$
'***---- Add other options here
setScriptResult HT_FAILURE
    Reset
    Exit sub
done:
logMsg "Script (rcvCall.sbl) is now finished running."
    Reset
End sub
```

**TABLE A-1    Measured Call Setup Time with One Type of IP-PSTN GWs**

| Call Progress Stage | Response Time (msec) | Functionality | Comments |
|---|---|---|---|
| First | 3920 to 5840 | Reaching the near-end GW | Voice heard |
| Second | 250 to 350 | Identification | Voice heard |
| Third | Not applicable | Authentication | Not applicable |
| Fourth or final | 7900 to 8900 | Ringing the called party's phone | Ring-back tone heard |

(i.e., the dial tone for the final stage of dialing is provided) to dial a destination phone number, provided that the authentication is successful.

The proposed technique can be used to measure the system response time at every stage of call setup. It is flexible, so that it can be easily modified to include new tone or a set of tones, or "voice begin" can be used at every stage. This method has been implemented using a set of scripts written in HVB language for measuring the call setup time using a number of commercially available IP telephony GWs. The results for one type of IP-PSTN GW are presented in Table A-1.

The same set of scripts can be used to measure the call progress/setup time for single-stage dialing as well. It is expected that the next-generation IP telephony GWs will allow single-stage dialing through a suitable routing/dialing plan in the central office switches and/or via digit manipulation at the ingress IP-PSTN GW.

## REFERENCES

1. Website of Hammer Technologies, www.hammer.com, 1999 (or http://www.empirix.com/empirix/voice+network+test/, 2001).
2. S. Gladstone, Testing Computer Telephony Systems and Networks, Flatiron Publishing, Inc., (now CMP Books) New York, 1996.

**Figure A-11**   A segment of the rcvCall.sbl routine written in HVB language. This script emulates a called party over an analog line (using the wink_start0 protocol). It answers the incoming call on the first ring. Then it plays a voice prompt for a prespecified length of time.

# APPENDIX B

# AUTOMATION OF CALL SETUP IN IP TELEPHONY FOR TESTS AND MEASUREMENTS[1]

In IP telephony, a call is usually established in multiple stages. In the first stage, an ingress or call-originating IP-PSTN GW is accessed. This is followed by a PIN-based caller authentication. Finally, the destination telephone number is entered. If the GWs have enough digital signal processing (DSP) channels and processing capacity, and the backbone (transport) network can support one T1 CAS port's worth of calls, we should be able to start 24 voice connection attempts *simultaneously*. The call-originating GW should be able to process all 24 connection requests. However, it appears that most of the currently available IP-PSTN GWs cannot handle all 24 connection requests simultaneously. Therefore, it is necessary to develop a method to determine the number of calls that can be started simultaneously. It is also necessary to determine the inter-call-burst time gap (in milliseconds or seconds) so that all 24 calls will be processed using the existing hardware and software configuration and capacity of the GW. This appendix describes techniques used to perform both of the above functions. They are implemented using Hammer' HVB language [1] (www.hammer.com) for testing some commercially available IP telephony GWs.

## INTRODUCTION

The emerging IP telephony GWs for enterprise networking applications usually support (a) one to four T1 ports per line card for interfacing to the PBX or PSTN network and (b) one or two auto-sensing 10/100 BT Ethernet ports for

---

[1] The ideas and viewpoints presented here belong solely to Bhumip Khasnabish, Massachusetts, USA.

152

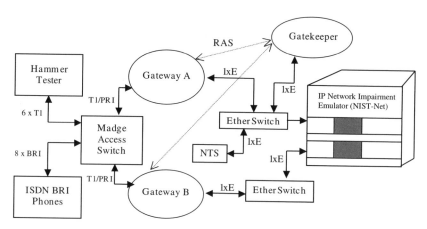

**Figure B-1**    A simple testbed for measuring the call setup performance of IP-PSTN GWs. E, Ethernet link; T1, CAS or PRI link; BRI, basic rate interface links. NTS stands for network time server; it provides timing information (clock) to the IP domain network elements such as IP-PSTN GWs, GK, and NIST-Net, and if needed, it can derive clocking information from a GPS receiver as well. A detailed description of the testbed is presented in Chapter 5.

interfacing to the IP network via a LAN. One possible configuration for supporting and testing black phone to black phone voice calls over an IP network is shown in Figure B-1. Generic information on computer telephony testing can be found in Reference 2. Note that a voice call originating from a PBX or PSTN network is considered to be the one from a black phone. Figure B-1 also shows where and how (i.e., using what types of links) various kinds of test equipment are connected to support different types of tests and measurements. A detailed description of the testbed can be found in Appendix A and Chapter 5.

The actual number of voice channels that can be used for voice conversation depends on the total number of DSP channels available in the GW. For example, if 16 DSP channels are supported per remote access service (RAS) card, a total of six ($= 96/16$) such cards would be needed to support the four T1 ports or $4 \times 24 = 96$ voice conversations. Alternatively, if 24 DSP channels are supported per RAS card, a total of four ($= 96/24$) such cards would be needed to support the four T1 ports or $4 \times 24 = 96$ voice conversations.

Now, assuming that we have enough DSP channels and processing capacity in the GW, and that backbone (transport) capacity is available to support one T1 CAS (or ISDN PRI) port's worth of voice conversation, we should be able to start 24 (or 23 for ISDN PRI) voice connection attempts simultaneously. The ingress (or call-originating) GW should be able to process all 24 connection requests. However, it appears that most of the first-generation IP-PSTN GWs cannot handle all 24 connection requests simultaneously. It is therefore necessary to develop a method to control the number of calls started simultaneously. At the same time, it is also possible to determine the amount of intercall-burst time gap (in milliseconds) needed so that all 24 calls will be processed with the existing hardware and software configuration and capacity of the GW.

A set of HVB [1] language-based scripts that can perform both of the above functions is presented in this appendix. The proposed technique and the results obtained are discussed. Some concluding remarks are presented next. The scripts for emulating the call-originating and call-terminating parties are also presented.

## THE PROPOSED TECHNIQUE

In this section, we first present a simple example, followed by a generic description of the proposed technique in the form of flowcharts. Next, another example is discussed. Implementations using HVB are then presented.

### A Simple Example

As mentioned earlier, with one set of T1 connection from the PSTN network, the ingress IP-PSTN GW should be able to support gracefully a maximum of 24 incoming call requests. However, because of limited DSP resources, very often most of the call attempts fail. To solve this problem, a precall wait can be added. The precall wait determines the amount of time the call-originating script will wait before actually making a connection attempt over a selected channel.

*In this example, only one call setup request is in progress at any point in time, and a precall wait time of 1 sec is used.*

Let us define a parameter called "waitTick" which controls the intercall time gap or precall waiting time, called "waitTime." In this simple example, only one call is started at a time, and a 1-sec intercall gap is added. In case of one T1, we have 24 channels to start the calls or connection requests. The call in the first channel starts at the same time that the script starts running on the channel, the call in channel 2 starts 1 sec after the script starts running, the call in channel 3 starts 2 sec after the script starts running, the call in channel 4 starts 3 sec after the script starts running, the call in channel 24 starts 23 sec after the time the script starts running in that channel, and so on. Therefore, the staging of calls is as shown in Figure B-2. An implementation of this example using HVB is shown in Figure B-3. In this example, the script that is making an outgoing call is running on channel 1 [indicated by the parameter chanId( )], and the script that will be receiving that call is running on channel 25, as indicated by the parameter "sendID." This depends on the system configuration and can be easily controlled.

Once the "placeCall" event has occurred on a channel, for one-stage dialing a ring tone would be expected from the called party's telephone. For multistage dialing, a dial tone or any specific predesigned tone or set of tones would be

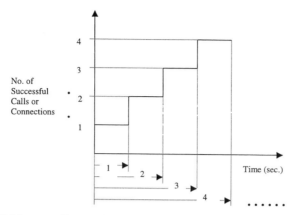

**Figure B-2**    Multistage call setup in IP telephony. One call setup request is in progress at any point in time. The intercall-burst (burst size = 1) time gap is 1 sec.

heard at the calling party's telephone, and the calling party can then proceed to the next stage of dialing.

### Description of the Technique

The proposed technique consists of the following three major procedures.

- Determining the maximum number of simultaneous incoming call requests that an ingress GW can handle so that a connection setup is allowed and a bidirectional conversation can proceed smoothly;

```
waitTick = 1
sendID = chanId( ) + 24

'***---- For dialing into a Hammer channel use the following ***
dialnumber = "655620" + str(sendID) + "#"
call startProtocol (HT_PROTO_WNK0,,)
waitTime = (sendID-25)* waitTick

'***---- Use HT_SECONDS for sec. level wait, HT_MILLISECONDS for
millisec level wait
pause waitTime, HT_SECONDS
logmsg "⇒ Pre-Call Wait is " & waitTime & " sec", HT_LOG_DEBUG

'***---- Enter the called party's number ***
logMsg "⇒ Placing call LU-GW4 B-to-A", HT_LOG_DEBUG
set event = placeCall (dialnumber,)
```

**Figure B-3**    A simple example of adding a precall waiting time of 1 sec between each call.

- Determining the intercall-burst time gap, which may vary from hundreds of milliseconds to a few seconds, depending on the GW and GK software, hardware, and processing capacity of the DSP modules of the GWs; and
- Developing a script or suite for determining the number of stages for setting up a prespecified number of calls in multiple stages. For example, if a GW can handle only four simultaneous incoming call requests, one would need six $(= 24/4)$ stages to set up 24 incoming call requests with a reasonable amount of intercall-burst time gap.

*Step 1:* In this step, the objective is to determine the maximum number of calls that can be started simultaneously so that the call attempts are successful. For one T1 connection between the PSTN network and the ingress/egress GW, it is possible to start the calls on all 24 channels of a T1 line of the Hammer tester [1,2] at the same time. Then one can visually monitor the status of the call requests in the Hammer tester's channel monitor. Usually, successful calls or connections are indicated by a green color and failed connections are indicated by red. If the connection requests are successful on all 24 channels, the ingress GW is capable of handling 24 simultaneous connection establishment requests. If not, one can either start with $(24 - 1)$ or 23 calls, then with $(24 - 2)$ or 22 calls, and so on until a point is reached at which all the connection attempts are successful. Alternatively, one can start with only one connection request, then increase it to two, three, four, and so on, and monitor the status of the connection requests in Hammer's monitor window to determine the maximum number (e.g., $n_1$) of calls that can be started simultaneously resulting in successful connections. A flowchart describing various phases of step 1 is presented in Figure B-4.

*Step 2:* Once step 1 is completed, one must determine experimentally the intercall-burst interval that allows establishment of connections on all 24 channels for one set of T1 connection to the IP-PSTN GWs from the Hammer tester. Our experiments show that we can start with 1 sec of intercall-burst time gap, and depending on the number of stages involved in call setup, we may need to increase it to a few seconds or decrease it to hundreds of milliseconds.

*Step 3:* The objective here is to utilize the results obtained in steps 1 and 2 in order to stagger the calls or connection requests so that the call attempts are successful on all 24 channels. If multiple (say $m$) T1 links are available from the PSTN network to the ingress GW, an attempt should be made to make successful calls on all $(m \times 24)$ channels. Note that the value of $n_1$ is determined in step 1 using the algorithm presented in Figure B-4. The value of the intercall-burst interval is determined in step 2. Figure B-5 presents various phases of step 3 in a flowchart.

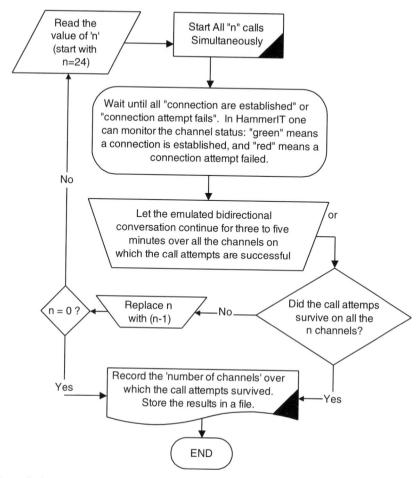

**Figure B-4**  Determination of the number of calls that can be started simultaneously so that successful connections are established to the called telephones.

## Another Example

In this example, attempts are made to start multiple calls simultaneously with an empirically determined size of the call bursts. Although it is possible to implement a nonuniform distribution of bursts, for the sake of simplicity we use uniform call bursts here.

Let us set the size of the call burst that represents the number of calls to start at the same time—that is, the parameter "no_of_simult_calls"—to 4. The intercall-burst interval, waitTick, is set to 3 sec. With this combination of no_of_simult_calls and waitTick, one would need six [= 24/4] stages and 15 [= 3 × ((24/4) − 1)] sec to start all 24 calls. If all of the calls are made and the call setup time is 5 sec, then successful establishment of all the connections

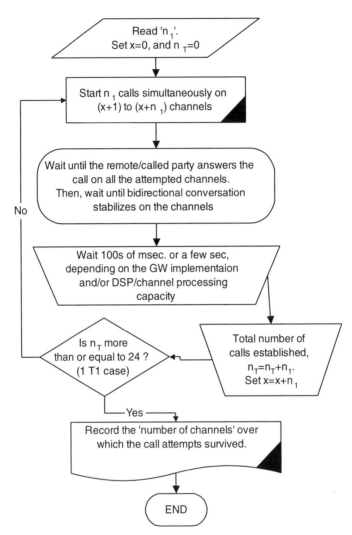

**Figure B-5**   Staggering of the calls or connection requests in order to establish 24 successful connections for one T1 link from the PSTN network to the GW. Note that the value of $n_1$ is determined by using the algorithm presented in Figure B-4.

would take approximately 20 sec. Figure B-6 presents an implementation of this type of *call staging* using the HVB language.

**An Implementation Using HVB Language**

In this section we present an implementation of the proposed technique using HVB language. The suite consists of the following two routines: "placeCall. sbl" and "receiveCall.sbl."

```
sendID = chanId( )        '*** chanId( ) is the Id of the channel on which the
                              script is running
no_of_simult_calls = 4    '*** set the size of call burst which is the no. of calls to
                              start at the same time
waitTick = 3              '*** wait for 3 sec. between each call bursts
'***---- 1st channel in Hammer Board#0 makes call to the 1st channel in Hammer
Board#1
sendID = sendID + 24
'***---- Telephone number of the destination channel for calls from Hammer
Board#0 to Board#1
dialnumber = "55720" + str(sendID) + "#"

'***---- Implementation of call staggering: calling channels: ch. 1–24, called
channels: ch. 25–48
for i = 1 to (24/no_of_simult_calls)
    for j = 1 to no_of_simult_calls
    k = (i – 1)*no_of_simult_calls + j
    waitTime(k) = (i – 1)*waitTick
next j, i

'***---- Inter-call-burst time gap for 24-channel calls from Hammer Board#0 to
the 24 channels in Board#1
logMsg " ⇒ wait for " & waitTime(sendID-24) & " seconds before making a
call", HT_LOG_DEBUG
pause waitTime(sendID-24), HT_SECONDS
```

**Figure B-6** A segment of the placeCall.sbl routine, which emulates the call staggering, with no_of_simult_calls as the number of calls started simultaneously and waitTick as the time gap in seconds. Between the $n$th and $n + 1$th call bursts.

The placeCall.sbl routine emulates a calling party. It implements the two telephony events needed to make a call setup request: placing a call and waiting for a response. Next, it is necessary to send a set of digits for user ID (optional) and wait for a response. Entering a set of digits for user authentication (optional) and waiting for a response follows. The final stage consists of sending the destination phone number and waiting for the ring-back tone. If the connection request is successful, a set of voice prompts must be played. The call is then released. Figures B-7a, B-7b, and B-7c present HVB implementation of the above functions. Figures B-8a and B-8b show implementation of various functions for detecting the call progress events and the beginning and end of voice prompt playout.

The receiveCall.sbl routine emulates a called party. It implements the telephony events—answering a call after a prespecified set of rings, playing of one or more voice prompts, and then releasing the call. The implementation is shown in Figures B-9a–c.

```
'***---- First Stage of Dialing: Dial the Ingress or Local Gateway's telephone
number
dial_local_gw = "5554005#"     '*** 4006 for egress or remote GW, for type-A
                                GW
bbb_det_window = 2             '*** set to 2 for 2 milliseconds detection window
after_bbb = 1500               '*** pause in milliseconds after detecting the
                                beepbeepbeep tone
call startProtocol (HT_PROTO_WNK0,,)
set event = placeCall (dial_local_gw,)     '* to place a call to the local or ingres
                                            GW

    eventType = event.type( )
    if eventType <> HTEVT_CALL_CONNECTED then goto TelError

'***---- Call progress time (t11) measurement for the first stage ----------------------
    set arSentEvent(1) = getCallEvent(HTCALL_DIGITS_SENT)
    set arRcvdEvent(1) = getCallEvent(HTCALL_REMOTE_ANSWERED)
    timeToAnswer = arRcvdEvent(1) − arSentEvent(1)
                                '*** timeToAnswer from placeCall to remote_
                                answered

'***---- Detection of the Beep Beep Beep tone -----------------------------------------------
if (det_bbb = 1) then
    toneDetParam.setVal HTPARM_TDET_TIME, bbb_det_window
    addTone 2,850,50,1477,50, toneDetParam
    set arRcvdEvent(2) = WaitForCPEvent(HTEVT_TONE_2_BEGIN)
    eventType = arRcvdEvent(2).type( )
        if (eventType <> HTEVT_TONE_2_BEGIN) AND (eventType <>
        HTEVT_DIGIT_BEGIN) then
        logMsg " !! Didn't detect Beep's Tone_2_Begin or HTEVT_DIGIT_
        BEGIN", HT_LOG_NORMAL
            goto TelError   '*** can collect CallFailedReason stats here
        end if

'***---- Call progress time (t12) measurement for the first stage (2nd half ) --------
    timeMillisec = arRcvdEvent(2) − arRcvdEvent(1)
                                '* time to answer from Remote_answered to BBB
                                tone

    removeTone(2)
end if
    pause after_bbb, HT_MILLISECONDS
```

**Figure B-7a**    A segment of the placeCall.sbl routine written in HVB language. This script emulates a calling party over an analog line (using the wink_start0 protocol). It dials a local or ingress IP-PSTN GW and then detects the specific tones before proceeding to the next stage of dialing for call setup. At every stage, it also measures the response time by subtracting the time of occurrence of the telephony events.

'* *Second Stage of Dialing: Dial the PIN number after the BeepBeepBeep tone is heard.*

'* Note: arSentEvent(1) is reused here for generic response time measurement*
dialPIN = "8935258221#"    '*** for type A GW only
dialTone_det_window = 15    '*** set to 15 for 15 milliseconds window
after_dialTone = 1000    '*** pause in milliseconds after detecting the
                              dialtone
if (pin_enabled = 1) then
   logMsg " ⇒ Sending DTMF digits for PIN number" &dialPIN, HT_LOG_
   NORMAL
   set arSentEvent(1) = sendDtmf(dialPIN,)
   eventType = arSentEvent(1).type( )
      if eventType <> HTEVT_TONES_DONE then
         logMsg "!! Unable to send PIN" goto TelError
                              '*** can collect CallFailedReason stats here
      end if
   logMsg " ⇒ PIN is entered; Dialtone should be heard immediately",
   HT_LOG_NORMAL
end if
'***---- *Detection of the standard US dialtone with tolerance(s)* ----------------------
if (det_dialTone_enabled = 1) then
   toneDetParam.setVal HTPARM_TDET_TIME, dialTone_det_window
   addTone 3,350,10,440,10, toneDetParam
   set arRcvdEvent(3) = WaitForCPEvent(HTEVT_TONE_3_BEGIN)
   eventType = arRcvdEvent(3).type( )
   if eventType <> HTEVT_TONE_3_BEGIN then
      logMsg "!! No dialtone is heard after PIN"
      goto TelError          '*** can collect CallFailedReason stats here
   end if

**Figure B-7a** *(Continued)*

For example, the placeCall.sbl script can be executed on all 24 channels of the first AG-T1 board, which is Board no. 0 of the Hammer tester (see Fig. B-1 for details). The receiveCall.sbl can be executed on all 24 channels of the second AG-T1 board (Board no. 1) of the same Hammer tester. The call staging routine presented in Figure B-6 can be used to establish all 24 calls.

## RESULTS

This section presents the results of the call setup automation experiments using Type A IP-PSTN GWs. The experimental setup is shown in Figure B-1. We used only one pair—one making outgoing calls and the other receiving incom-

```
'***---- Call progress time (t2) measurement for the second stage --------------------
timeMillisec = arRcvdEvent(3) − arSentEvent(1)
  logMsg " ⇒ " & timeMillisec & " ms ... to hear dialTone after PIN is
  entered", HT_LOG_NORMAL
  removeTone(3)
end if
pause after_dialTone, HT_MILLISECONDS

'***---- Third Stage of Dialing: Dial the destination telephone number after the
dialtone is heard.*
set arSentEvent(3) = sendDtmf(dialnumber,)
  eventType = arSentEvent(3).type( )
  if eventType <> HTEVT_TONES_DONE then
    logMsg "!! Unable to complete the detination call" goto TelError
  end if

'***---- Detection of the standard US ringtone with tolerance(s) ----------------------
  toneDetParam.setVal HTPARM_TDET_TIME, ringTone_det_window
  addTone 2,440,10,480,10, toneDetParam
  set arRcvdEvent(4) = WaitForCPEvent(HTEVT_TONE_2_BEGIN)
  eventType = arRcvdEvent(4).type( )
  if eventType <> HTEVT_TONE_2_BEGIN then
    logMsg "!! No ringtone is detected, try 2 ms detection window." goto
    TelError
  end if

'***---- Call progress time (t3) measurement for the third stage ----------------------
  timeMillisec = arRcvdEvent(4) − arSentEvent(3)
  logMsg " ⇒ " &timeMillisec & " ms ... to hear Ring-back tone at calling
  end", HT_LOG_NORMAL
  removeTone(2)
pause after_ringTone, HT_SECONDS
Do While x < repeatTimes
ClearDigits

'***---- Now start playing voice prompts -----------------------------------------------
promptID = "m" &x &".WAV"
  logMsg "⇒ Playing " & promptID, HT_LOG_DEBUG
  set event = playPrompt(promptID, HT_ENCODE_PCM8M16, 10000,)
  eventType = event.type( )
  if eventType <> HTEVT_PLAY_DONE then goto TelError
pause 12, HT_SECONDS   '*** the pause after playout of voice prompt
promptID = "voipman1p" &x &".pcm"
  logMsg " ⇒ Playing " & promptID, HT_LOG_DEBUG
  set event = playPrompt(promptID, HT_ENCODE_PCM8M16, 10000,)
  eventType = event.type( )
  if eventType <> HTEVT_PLAY_DONE then goto TelError
pause 9, HT_SECONDS   '*** the pause after playout of voice prompt
x = x + 1              '*** playPrompt has been repeated "x" times
Loop
```

```
set event = releaseCall( )
eventType = event.type( )
if (eventType <> HTEVT_CALL_RELEASED) AND (eventType <>
HTEVT_CALL_DISCONNECTED)
    then goto TelError
setScriptResult HT_SUCCESS
stopProtocol
logMsg "Script (placeCall.sbl) is now finished running."
Reset
End sub
```

**Figure B-7c**   This segment of the placeCall.sbl routine emulates the release of a call, displays the message, and then completes execution of the script.

ing calls—of T1 links between the Hammer tester and the Madge Access Switch.

A set of HVB scripts has been developed to implement the algorithms presented in the previous section. For one set of T1 links between Hammer and Madge, 24 calls can now be successfully established from the ingress GW, A, through the Madge switch, the IP network, and the egress GW, B. Using the algorithm presented in Figure B-4, we find that the maximum number of calls that can be started simultaneously to achieve successful call establishment is *four*. Next, it is necessary to determine the value of the waitTick (see Fig. B-6 for details) parameter so that the call bursts are sufficiently spaced in the time domain. This ensures that the call processing–related stress on the DSPs and the related CPU is distributed (i.e., spread over time) such that the call attempts are neither rejected nor failed. Using the algorithm presented in Figure B-5, we find that the waitTick parameter needs to be set at 6 sec to ensure establishment of all 24 calls. These results and the call setup times for various values of waitTick are presented in Table B-1.

Note that the number of calls survived does not increase uniformly with an increase in the value of the waitTick parameter. This can be attributed to the fact that the call processing capability of the DSP resources is depleted more rapidly than the rate at which the utilization of DSP resources increases. This seems to be true for many of the commercially available first-generation IP-PSTN GWs.

**Figure B-7b**   A segment of the placeCall.sbl routine, which emulates the second and third stages of dialing. It measures the response time by subtracting the time of occurrence of the telephony events. It then emulates playing of a set of voice prompts with pauses to accommodate play-out of the called party's voice prompts. Finally, it emulates the release of a call, displays the message, and then completes execution of the script.

```
'* Implementation of the wait for call progress event ( WaitForCPEvent) as a
telephony event
'* It can detect both tone (e.g., digit_begin) and other call progress events        *
Function WaitForCPEvent(cpevent as double) as telEvent
dim event        as telEvent
dim i, done      as integer
dim cpParams     as parmCallProg
dim eventType    as double

  i = 0
  done = 0
  while done = 0
    set event = getNextEvent(20000)
    eventType = event.type( )
    if(eventType <> HTEVT_TIMED_OUT) AND (eventType <> HTEVT
    _UNINITIALIZED) then
      event.eventText HT_eventStr
      logMsg " ⇒ Event: " & HT_eventStr, HT_LOG_NORMAL
    end if           '*** usually the TIMED_OUT onewill be the one to come up

    i = i + 1
    if i = 7 then    '*** number of times (e.g., 6 sec/time) for detection
      done = 1
      logMsg " ⇒ Detection Failed in loop for " &i &"times!", HT_LOG_
      NORMAL
    end if
    if (eventType = cpevent) or (eventType = HTEVT_DIGIT_BEGIN) or
    (eventType = HTEVT_CP_DONE) then
      done = 1
    end if
  wend

  ' if eventType <> HTEVT_CP_DONE then stopCallProgress
  set WaitForCPEvent = event

End Function
```

**Figure B-8a**   This segment of the placeCall.sbl routine emulates the wait for call prog-ress event function. The objective is to exit from the loop when the required telephony event occurs.

```
'*****************************************************************
'* Implementation of the Function WaitForVoiceEnd event as an integer    *
'* It determines the time it takes to hear voice coming over the line by using  *
'* call progress analysis to determine when voice is heard. It RETURNS   *
'* "−1" on error or "1" on voice_end.                              *
'*****************************************************************
Function WaitForVoiceEnd(cptimeout as double) as integer

dim voiceHeard   as integer          '*** used to determine called party responds
dim voiceEnd     as integer
dim mycptime     as parmCallProg
mycptime.SetVal HTPARM_CP_TIMEOUT, cptimeout
startCallProgress   mycptime          '*** should be longer than voice length

   voiceHeard = 0
   do
      set event = getNextEvent(20000)
      eventType = event.type( )
      if (eventType = HTEVT_CP_VOICE) then
         startStopWatch(1)
         voiceHeard = 1
         logmsg " ⇒ Voice Begin . ."
      elseif (eventType = HTEVT_CP_DONE) then

'***---- Callprogress analysis stopped before voice was heard, this is a problem;
return −1
         WaitForVoiceEnd = −1
         logmsg "⇒ Wait for voice begin: CallProgress Timeout in mycptime="
         &cptimeout &" seconds"
         exit function
      end if
   loop while voiceHeard = 0

StartVoiceEnd:    '*** voice has been heard, now wait till voice ends.
   voiceEnd = 0
   do
      set event = getNextEvent(20000)
      eventType = event.type( )
      if (event.value( ) = HT_CP_VOICE_END) then
         logMsg " ⇒ BK1 = The voice heard is about " & readstopwatch(1) &" ms
         long", HT_LOG_NORMAL
         voiceEnd = 1
         logmsg " ⇒ Voice End"
      elseif (eventType = HTEVT_CP_DONE) then
```

**Figure B-8b**  This segment of the placeCall.sbl routine emulates the wait for call progress event function. The objective is to exit from the loop with an error message or when the play-out of the voice prompt ends.

```
'***---- Callprogress analysis stopped before voice was heard, this is a problem;
return −1
WaitForVoiceEnd = −1
        logmsg "⇒ Wait for voice end: CallProgress Timeout in ? seconds"
        exit function
      end if
   loop while voiceEnd = 0
   WaitForVoiceEnd = 1

stopCallProgress   '*** stop call progress analysis
End Function
```

**Figure B-8b** *(Continued)*

## CONCLUSIONS

A method to establish a large number of IP telephony calls automatically
has been presented. It consists of determining the *number* of calls that can be
started simultaneously and the *intercall-burst interval* (in milliseconds or seconds). The intercall-burst interval enables processing of all incoming call
requests in multiple stages using the existing hardware and software configuration and capacity of the GW and GK. An implementation of the proposed
methods using Hammer's (www.hammer.com) HVB language for testing some
commercially available IP telephony GWs is also included in this appendix.
Both single-stage and multistage call setup can be handled efficiently using the
proposed technique.

```
call startProtocol (HT_PROTO_WNK0,,)
logMsg "⇒ Waiting for an incoming call...", HT_LOG_DEBUG
   set event = waitForCall(−1)
   eventType = event.type( )
   if eventType <> HTEVT_INCOMING_CALL then goto TelError
logMsg " ⇒ Receiving or Answering an incoming call...", HT_LOG_DEBUG
   set event = answerCall(2)
   eventType = event.type( )
   if eventType <> HTEVT_CALL_CONNECTED then goto TelError
```

**Figure B-9a**   A segment of the receiveCall.sbl routine written in HVB language. This
script emulates a called party over an analog line (using the wink_start0 protocol). It
answers the incoming call after two rings.

```
x = 1
repeatTimes = 6
Do While x < repeatTimes
  clearDigits
  promptID = "f" &x &".WAV"
  logMsg "⇒ Playing " & promptID, HT_LOG_DEBUG
  set event = playPrompt(promptID, HT_ENCODE_PCM8M16, 10000,)
  eventType = event.type( )
  if eventType <> HTEVT_PLAY_DONE then goto TelError

pause 12, HT_SECONDS   '*** the pause after playout of voice prompt

  promptID = "voipwom1p" &x &".pcm"
  logMsg "⇒ Playing " & promptID, HT_LOG_DEBUG
  set event = playPrompt(promptID, HT_ENCODE_PCM8M16, 10000,)
  eventType = event.type( )
  if eventType <> HTEVT_PLAY_DONE then goto TelError

pause 9, HT_SECONDS   '*** the pause after playout of voice prompt
x = x + 1
Loop
```

**Figure B-9b**   This segment of the receiveCall.sbl routine emulates playing of a set of voice prompts with pauses to accommodate play-out of the calling party's voice prompts. This is repeated six times with different voice prompts, numbered 1 through 6.

```
set event = releaseCall( )
  eventType = event.type( )
  if eventType <> HTEVT_CALL_RELEASED then goto TelError
logMsg "⇒ Call released.", HT_LOG_DEBUG
setScriptResult HT_SUCCESS
stopProtocol
logMsg "Script (receiveCall.sbl) is now finished running."
Reset
End sub
```

**Figure B-9c**   This segment of the receiveCall.sbl routine emulates release of a call, displays that message, and then completes execution of the script.

**TABLE B-1   Call Setup Performance for a Commercially Available IP-PSTN GW**

| Intercall Burst Time Gap or the waitTick Parameter (sec) | No. of Calls Survived | Call Setup Time (sec) | Delay Vector (sec) |
|---|---|---|---|
| 0 | 04 | 3.9 to 4.70 | $\{0,0,0,0,0,0\}$ |
| 1 | 08 | 3.4 to 10.2 | $\{0,1,2,3,4,5\}$ |
| 2 | 12 | 3.5 to 10.2 | $\{0,2,4,6,8,10\}$ |
| *3* | *15* | *3.5 to 10.2* | *$\{0,3,6,9,12,15\}$* |
| 4 | 18 | 3.5 to 10.2 | $\{0,4,8,12,16,20\}$ |
| 5 | 19 | 3.5 to 10.2 | $\{0,5,10,15,20,25\}$ |
| 6 | 24 | 3.5 to 10.2 | $\{0,6,12,18,24,30\}$ |

## REFERENCES

1. Website of Hammer Technologies, www.hammer.com, 1999 (or http://www.empirix.com/empirix/voice+network+test/, 2001).

2. S. Gladstone, Testing Computer Telephony Systems and Networks, Flatiron Publishing, Inc., (now CMP Books) New York, 1996.

# APPENDIX C

# EVALUATION OF VoIP SERVICES[1]

This appendix presents experimental analyses of the media path's QoS in IP-based telephony. The media path or bearer path is used to transfer information during a session. In an IP-based network (e.g., the Internet), the media path is a routed path and can be used to transmit both voice and tones in real time. We analyze the characteristics of the media path by transmitting (a) a voice signal, (b) a DTMF (dual tone multiple frequency) signal, and (c) voice and DTMF signals. We use the Hammer tester's implementation [1] of ITU-T's perceptual speech quality measurement (PSQM) score [2] based voice quality measurement technique to evaluate the quality of speech transmission over an IP network. Other techniques include determining the PSQM+, PAMS, and PESQ scores (these terms are defined in the Glossary) for voice transmission. For assessing the quality of DTMF transmission, we use a score of 1 for correct transmission and 0 for severely delayed and/or incorrect transmission.

## INTRODUCTION

In traditional telephone networks or PSTN, voice transmission services are delivered using the traditional circuit-switching technology. This is a very robust technology, but it is neither flexible nor cost-effective. Therefore, other switching methods such as packet switching need to be explored. The emerging telecom companies are building packet—mostly IP or IP-based—network infrastructures [3] to provide a variety of packet-based services including

---

[1] The ideas and viewpoints presented here belong solely to Bhumip Khasnabish, Massachusetts, USA.

**169**

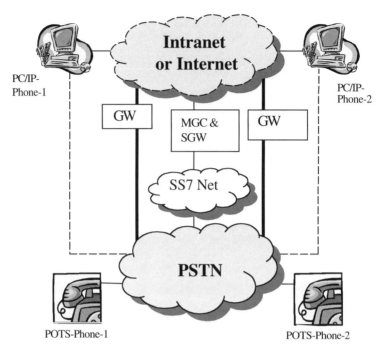

**Figure C-1** Evolving telephone network.

enhanced services such as VoIP, fax over IP, messaging over IP, and so on using the same network. Figure C-1 explains the evolving scenario. The IP-PSTN GWs facilitate transmission of a TDM-formatted (or circuit-switched) voice signal over an IP-based network (an Intranet or the Internet). The media gateway controller (MGC) controls the GWs and the calls that are routed through them, and the SS7 signaling gateway (SG) interprets PSTN domain signaling messages (i.e., SS7 messages) in the IP domain and vice versa. A connection establishment request from POTS-Phone-1 (plain old telephone system) to POTS-Phone-2 can be routed through one of the two networks: (a) from PSTN to PSTN over a PSTN network or (b) from PSTN through the Internet to the PSTN. Also, in order to establish a connection from PC/IP-Phone-1 to PC/IP-Phone-2, any one of the following four paths can be used:

a. From Internet to Internet (worse performance, but inexpensive or free)
b. From PSTN to Internet to PSTN (desirable)
c. From Internet to PSTN to Internet (not desirable)
d. From PSTN to PSTN (best performance but expensive)

These scenarios reveal that different routes can be used to establish a communication session between the two endpoints (phones/PCs), depending on the desired quality of service requirements. The same flexibility can be used to avoid network congestion during heavy utilization of one or more of the paths as well.

In today's telephone networks, when a user makes a call from POTS-Phone-1 to POTS-Phone-2, the call can be routed through either the Internet, an Intranet, or the PSTN, depending upon the calling plan one has, the price one pays, or the network routing, which may depend on the availability of network resources.

In PSTN-based routing, a direct or transparent connection is established from POTS-Phone-1 to POTS-Phone-2. However, if the call is routed through the Internet, it uses a connectionless circuit. The E.164 telephone address is translated into the IP address through the MGC. Then the call is routed to the IP address of the MGW that is serving the destination phone (POTS-Phone-2).

The problem with the IP network (e.g., the Internet) is that it is packet based, and it is neither very reliable nor robust for sessions or services such as real-time voice communications. For example, some voice packets may arrive sooner than others, causing out-of-order delivery, which may result in impaired voice communications. However, the IP-based network offer flexible inter-working, rapid creation and marketing of novel services, and low-cost voice transmission. The reason for interworking between the Internet and PSTN networks is that most of the large telecom companies have billions of dollars invested in the PSTN infrastructures, and they cannot afford to write off these infrastructures quickly. Interworking between the packet and circuit-based networks can help the existing service providers get a full return on their investment in the PSTN networks.

## CONFIGURATION OF THE TESTBED

The configuration diagram of the testbed is shown in Figure C-2 (described in detail in Chapter 5). The Hammer tester is used for generating and analyzing

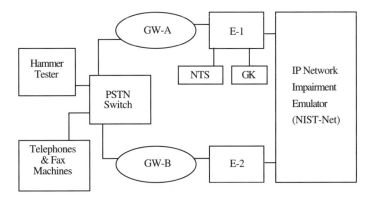

**Figure C-2**  Configuration of a testbed for measuring the quality of speech and DTMF signal transmission over an IP network.

the emulated PSTN phone to PSTN phone calls. The Madge Access Switch emulates a small PSTN central office (CO) switch. Madge can provide one or more T1-CAS and/or T1-PRI connections to the PSTN interfaces of the VoIP or IP-PSTN gateways (GW-A and GW-B) under test. The Intranet (or local Internet) of the testbed consists of two Ethernet switches (E-1 and E-2), and an IP network impairment emulator called NIST-Net (http://snad.ncsl.nist.gov/itg/nistnet/). NIST-Net is a PC-based system consisting of the Linux operating system. VoIP GW-A and GW-B are the near-end (ingress or call-originating) and far-end (egress or call-terminating) GWs. The gatekeeper (GK) of the testbed performs registration, administration/authentication, and status (RAS) monitoring functions when a call is registered. The network time server (NTS) provides timing information (clock) to the IP domain network elements such as IP-PSTN GWs, GK, and NIST-Net. If necessary, it can derive clocking information from a GPS receiver as well.

## MODEL OF A TEST CALL

In a typical telephone conversation session, there are two or more interacting players: for example, a calling party, a called party, an interactive voice response (IVR) unit, and so on. In the Hammer tester, a conversation is emulated by using a test suite that consists of at least two HVB scripts; one emulates a caller and the other emulates a called party, with communications occurring over the line or channel (over the Intranet) under test. Figure C-3 shows a ladder diagram of the sequence of interactions between the two HVB scripts playing the roles of caller and call receiver. Note that the sequence of play prompt and pause can be executed a number of times in order to increase the length of the emulated call.

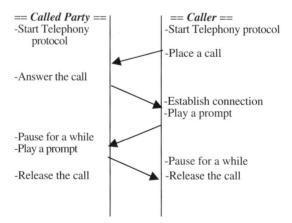

**Figure C-3**  Sequence of interactions between the calling and called parties during a typical telephone conversation.

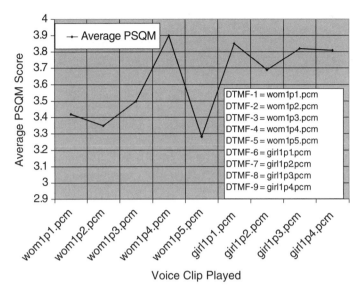

**Figure C-4**    Average PSQM scores for different types of voice samples.

## BASE CASE EXPERIMENTS AND RESULTS

In this case, the PSQM scores (0: best match or a good channel or transmission; ~6.5: worst match or a bad channel or transmission) are measured using the Hammer tester for a set of voice samples separately on both sides—sending and receiving—of the channel over the idle IP network without any impairment. Afterward, the average value is computed and a graph is plotted for the average PSQM value against the voice sample being played. The results are shown in Figure C-4.

## RESULTS OF EXPERIMENT 1

The effects of three different types of impairments, that is, packet loss, network delay, and jitter, are measured using four different voice clips—man1p2.pcm, boy1p2.pcm, girl1p2.pcm, and wom1p2.pcm—each playing the same sentence or message. The impairments are introduced separately, that is, only one type of impairment is introduced at any point in time using the NIST-Net. The results are as presented in Figures C-5, C-6, and C-7. It is clear that both packet loss and delay jitter significantly impair voice quality compared with network delay. As the value of delay jitter increases, the call-progress tones and speech signal become unintelligible. Also, the higher the value of network delay, the more difficult it becomes to establish a call or connection. This can be attributed to expiration of various timers during the call setup stage.

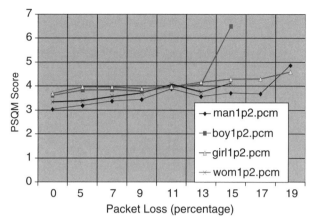

**Figure C-5**  Variation of the PSQM score with packet loss.

## RESULTS OF EXPERIMENT 2

In this experiment, the effects of three different impairments—packet loss, delay jitter, and network delay—are measured on the combination of voice and DTMF signal transmission. Each DTMF digit is used to represent a voice clip in the Hammer script. The correlation between the DTMF and the voice clip is as presented in the legend of Figure C-4. The effects of network impairments on voice signal transmission are measured using the PSQM score. In DTMF digit transmission, if it is recognized correctly at the other end of the channel,

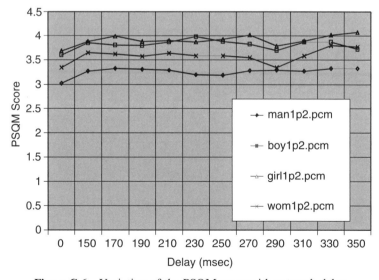

**Figure C-6**  Variation of the PSQM score with network delay.

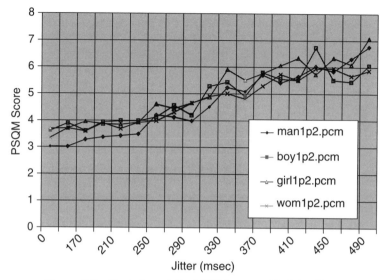

**Figure C-7**   Variation of the PSQM score with delay jitter.

the appropriate voice clip is played (score = 1); otherwise, either no voice clip is played or an incorrect voice clip is played (score = 0). The final score for DTMF digit transmission is computed by averaging the scores of all possible (i.e., one to nine) DTMF digit transmissions.

The emulated caller (Fig. C-3) *randomly* selects a set of DTMF digits and sends them over the preset transmission channel one after the other, with a predetermined amount of pause between them. A random number generator

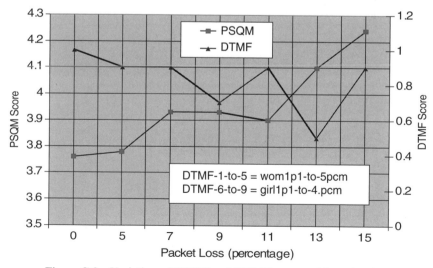

**Figure C-8**   Variation of PSQM and DTMF scores with packet loss.

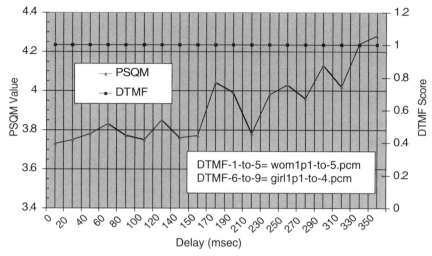

**Figure C-9**   Variation of the PSQM value and the DTMF score with network delay.

is used in the caller Hammer script to achieve this. The emulated called party plays the voice clips corresponding to the received DTMF digits (Fig. C-4). The call duration is set at approximately 5 min.

At the end of the experiment, sample averages are computed for both PSQM and DTMF scores, and the results are plotted on a graph against the different types of impairments. The results are plotted in Figures C-8, C-9, and C-10. It is clear that packet loss and delay jitter network impairments have the most

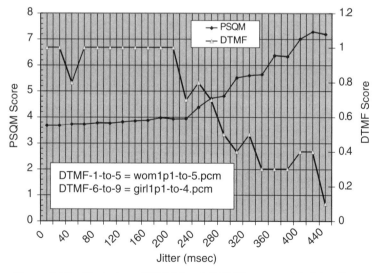

**Figure C-10**   Variation of PSQM and DTMF scores with delay jitter.

significant impact on the average PSQM score and the average DTMF transmission score values. The average DTMF score seems to remain unaffected until the delay jitter value reaches approximately 200 msec. Once again, the impairments are introduced by NIST-Net one at a time; combinations of two or more impairments are not used. The DTMF digits are generated randomly to simulate real-world application scenarios such as a business transaction or a banking application, where the user has to go through a few different stages or phases in order to complete a transaction.

## CONCLUSIONS

The experimental results presented in this appendix reveal that transmission of both voice and DTMF signals over IP networks is most affected by network impairments such as packet loss and delay jitter. Network delay seems to have the least impact on voice and DTMF transmission. Moreover, DTMF transmission does not seem to be affected by network delay. During experiments, it has been found that call establishment attempts sometimes fail repeatedly. This can be attributed to factors such as high values of delay jitter, packet loss, and network delay. During this study, only one network impairment is introduced at a time. Therefore, in future studies it is very important to perform these experiments using a mixture of different types of impairments.

The results obtained from this research can be used to develop threshold points for IP network operations. This can be very helpful for maintaining a better quality of (real-time) voice transmission and preventing service outage.

## REFERENCES

1. Website of Hammer Technologies, www.hammer.com, 1999 (or http://www.empirix. com/empirix/voice+network+test/, 2001).
2. P.861 Recommendation, Objective Quality Measurement of Telephone-Band (300–3400 Hz) Speech Codecs, ITU-T, Geneva, 1998.
3. D. Minoli and A. Schmidt, Internet Architectures, Wiley Computer Publishing, New York, NY, USA, 1999.

# GLOSSARY OF ACRONYMS AND TERMS[1]

**AAA**  Authentication, authorization, and accounting; a suite of network security services that provides a major framework through which access control can be implemented on any access server.

**AAL**  ATM (defined later) Adaptation Layer; the functions of translating application layer data or information into size and format of ATM cells. AAL-1 through AAL-5 have been defined; AAL-1 is used for constant bit rate and circuit emulation services for transmission of real-time voice and video, AAL-5 is used for variable bit rate connection-oriented and connection-less services (e.g., for IP over ATM).

**ACD**  Automatic call distributors; ACDs are designed to handle incoming phone calls or to make outgoing calls. Using ANI/DNIS, information collected via IVR, and by looking in a database (local or distributed, for intelligent call routing) ACDs can answer an incoming call by playing a pre-recorded message or can put the caller to the 'queue' from which a call agent (or an operator) is answering the incoming calls.

**ACELP**  Algebraic-code-excited linear-prediction; a technique utilized by G.723 voice coding scheme to generate 5.3 Kbps streams of data.

**ACM**  Address complete message; an ISUP message for telephone call setup and control using the SS7 network. This message is used to indicate the completion of address information.

---

[1] As the computer telephony integration (CTI) and voice over IP (VoIP) technologies evolve, many new acronyms and terms will be introduced; up-to-date information on these can be found at the following websites: www.ietf.org, www.iptelephony.org, www.itu.int, www.w3c.org, and www. sipforum.org.

**ADSL**   Asymmetric DSL; this refers to a version of DSL where the upstream and down stream data rates are asymmetric; G.Lite is a popular version of ADSL that delivers a data rate of 1.5 Mbps downstream (to home), and 640 kbps upstream (from home, toward the ISP or Telco).

**AGW**   Access gateway (*see GW*); an IP-PSTN gateway that supports one or more Ethernet (10/100 BaseT, gigabit Ethernet, etc.) interfaces on the packet side and one or more PSTN access lines (multiple DS0s, T1-CAS/PRI, etc.) on the PSTN side.

**AIN/IN**   Advanced intelligent network/intelligent network; this refers to a virtually separate and distributed telephone call processing architecture using service control point (SCP or remote control node), service switching point (SSP or enhanced CO), and intelligent peripheral (IP) or a dedicated service node as network elements. The objective is to achieve vendor and platform independence to rapidly introduce novel services. Some extensions to SS7 signaling standard was also developed to provide a framework for interaction of SSP, SCP, and IP components. For example, the protocol, like intelligent networking application part (INAP) defines a number of triggers needed to complete a particular service. Assembling the INAP operations into different sequences can create new services.

**A-Law**   An ITU-T specification for logarithmic conversion between analog and digital signals for pulse code modulation (PCM) technique in G.711 coding with the objective of improving the noise performance; used mainly in Europe and many other countries ($\mu$-Law is used in North America and Japan; *see $\mu$-Law*).

**ANI/DNIS**   Automatic number identification/dialed number identification System; ANI/DNIS is a telephone call processing feature which allows identification of the number originally dialed by a caller, and is widely used for routing toll-free (like 800, 888, etc.) calls, identifying appropriate call agent to answer an incoming call in a call center, etc.

**ANM**   Answer message; an ISUP message for telephone call setup and control using the SS7 network. This message is used to indicate answer from the called party so that a bi-directional connection (or circuit) can be established.

**ANSI**   American national standards institute; ANSI adapts the standards developed by other National and International Standards committees for use within the United States (see www.ansi.org).

**AMA**   Automated message accounting; this refers to a Telcordia (formerly Bellcore) recommended (GR-508-CORE) format for collecting PSTN call related general management and accounting information for billing and accounting purposes.

**API**   Application programming interface; an interface that software developers can use to write innovative applications programs for emerging services (e.g., *see JAIN*).

**Application Server**   A server hosting application that can be invoked by end users, such as the e-mail server, centrex feature server, unified message server, instant message server, and IVR server.

**ASP**   Application service provider; this refers to an ISP or advanced Telecom service provider who provides monthly-fee based access to advanced applications and services over the Internet (dial-up, DSL, or T1 link) to Enterprise or residential customers.

**ATM**   Asynchronous transfer mode; this refers to a packet switching or transfer technology which supports a variety of service-specific segmentation and reassembly (SAR) of information for adaptation—using a separate ATM adaptation layer or AAL—to transfer information using fixed-size (53 Bytes; 5 Byte header and 48 byte information) packets called cells. The transfer mode is asynchronous because the information from an individual user or application does not need to appear in periodic or synchronous fashion for transmission.

**ATM Forum**   This refers to an international organization of ATM based service providers and equipment manufacturers, which develops standards and specifications (available at www.atmforum.com/standards/approved.html) for ATM products and their Interoperability.

**BAF**   Billing AMA format; a Telcordia (formerly Bellcore) recommended (GR-1100-CORE) format for collecting PSTN call-related management and accounting information for billing purposes.

**BGP**   Border gateway protocol; an IETF protocol (see, e.g., RFC 1654) that defines routing in an inter-autonomous system (AS) by exchanging network reachability information with other BGP systems.

**BHCA**   Busy hour call attempt; a measure of the telephone switching system's performance. In VoIP, because of the distributed nature of the architecture, this may not be an adequate measure of the call-handling performance.

**BRI**   Basic rate interface; the ISDN BRI interface consists of two B channels (each 64 Kbps) and one data or signaling channel of 16 Kbps. Thus, one BRI link becomes 144 Kbps channel.

**BICC**   Bearer independent call control; an ITU-T call control protocol (Q.1901, June 2000) for adapting ISUP messages to support narrowband ISDN services independently of the signaling and transmission technologies.

**Busy Hour**   A time period during which the largest number of telephone call setup requests arrives; this knowledge helps telephone companies design the call-handling capacity of their PSTN switches.

**CALEA**   Communications assistance for law enforcement act; CALEA requires that the Telecom service providers comply with authorized surveillance of their communications and service facilities (see www.fcc.gov/calea/ for further details).

**CAS**   Channel associated signaling; the method of signaling, which utilizes one or more bits from the media (or voice) channel to indicate the state of the channel (or circuit).

**CASP**  Communications ASP; this refers to an ISP or advanced Telecom service provider who provides monthly-fee based access to advanced communications services—like unified messaging, Web based conferencing, follow-me/find-me services, etc.—over the Internet (using DSL or T1 link) to Enterprise or residential customers.

**CC**  Call controller; this refers to a server or packet router or a combination of both which controls and/or mediates setup and teardown of a VoIP call irrespective of the underlying protocol (H.323, SIP, MGCP, H.248/Megaco, etc.). In MGCP, a call controller is referred to as call agent (**CA**), in H.323 a call controller is referred to as gatekeeper (**GK**), and in H.248/Megaco, a call controller is referred to as a media gateway controller (**MGC**), and so on.

**CDR**  Call detail record; information related to a call, which usually includes data on calling and called parties, length of the call, call termination or drop reason code, and so on. The CDR can be used to generate billing records, to generate call patterns and statistics for network capacity planning, and to diagnose call-handling problems of the system.

**Centrex**  Central office exchange; this refers to a set of advanced and automatic (or pre-programmed button-based) call control and call distribution features which Businesses and high-end residential customers subscribe from their Telecom Service providers (usually software based, and hosted and maintained in the central office or CO switch in PSTN Network).

**CELP**  Code excited linear prediction; a technique commonly utilized in low-bit rate voice coding algorithms like G.723 and G.729.

**CGI**  Common gateway interface; the standard method for passing data or information from server to application program, and vice versa in a trusted environment.

**CIC**  Circuit identification code; a decimal digit string–based identifier in the SS7 protocol (MTP level 3) header used to identify the selected trunk for call establishment. CIC is also used to identify the interexchange carrier (IEC) lines for routing inter-LATA calls; in that scenario CIC stands for *carrier identification code*.

**CLASS**  Custom local area signaling services; this refers to a set of call control features—like caller ID, call forwarding, call waiting, automatic call back, selective call acceptance/rejection/forwarding, distinctive ringing, etc.—that are available from the local telephone switch or end office switch or CLASS-5 switch.

**CLEC**  Competitive local exchange carrier; a local communication (primarily access) service provider that offers voice telephony services in a LATA using leased or owned network and switching devices.

**CM**  Cable modem, the modulation-demodulation (modem) device of the customer's premise equipment to facilitate voice and data communications over CATV network. CM is a part of the DOCSIS (defined later) standard.

**CMTS** Cable modem termination system, the modem termination part—routers and bridges at cable head end—of the DOCSIS (defined later) standard.

**CNG** Comfort noise generation; generating background white (or Gaussian) noise locally and feeding it to the listening device. CNG is needed when silence suppression is used so that the silence signal from a talker does not need to be transmitted over the network. However, silence suppression may give the false impression that (a) the transmission quality is bad, (b) a call was disconnected, (c) voice packets are lost in transit, or other problems. Therefore, CNG is needed to complement the use of SAD or VAD.

**CO** Central office or end office telephone switch that commonly originates, terminates, or switches traditional voice telephony calls.

**CODEC** or **codec** coder-decoder; a coder performs sampling, quantizing, and associated processing of analog (e.g., speech/voice, video) signals with the objective of digitizing them; the decoder performs the reverse process to regenerate the analog signals. G.711, G.723, and G.729 are three common ITU-T-recommended voice coding standards.

**COPS** Common open policy service; this refers to an IETF protocol (RFC 2748, RFC 2749, RFC 2753, RFC 2940, RFC 3084) which describes a client-server model for enforcing policy based management of communication resources for guaranteeing application level quality of service.

**CoS** Class of service; a technique for classifying different traffic flows into a number of categories and applying a particular QoS for transmission of each of these categories of flow.

**CPE** Customer premise equipment; this refers to the terminal equipment or end-user device which reside within the customer's premise, and generate and/or consume real-time and non-real-time audio, video, and data information; e.g., a multimedia capable PC connected to the Internet via a PSTN modem or an IAD (defined later).

**CPL** Call processing language; a text- or script-based simple language that describes how the IP telephony call setup messages should be processed (see e.g., IETF's RFC 2824 for further details).

**CPS** Calls per second; the number of call setup requests that arrive at a switch (in PSTN) or a CC (in VoIP).

**CRTP** Compressed real-time transport protocol; an IETF specification (RFC 2508) for compressing IP/UDP/RTP headers (12 to 40 bytes) into 2 to 4 bytes.

**CS-ACELP** Conjugate structure algebraic code excited linear prediction; an algorithmic compression of digitized speech using human vocal tract model. This method is utilized in G.729 coding of voice signal to generate a bit stream of 8 Kbps of speed.

**CSR** Customer service representative; a live or automated agent in a call center to help resolve service related issues to customers over telephone line or web interface or both.

**Dejitter Buffer** or **Dynamic (Delay) Jitter Compensation Buffer**    This refers to a memory segment or buffer which temporarily accumulates the incoming voice packets for assembling them with evenly spaced time intervals, and subsequently delivering them to the voice play out buffer. The objective is to minimize the effect of delay jitter or variations on voice quality.

**Delay** or **VED**    Delay or voice envelop delay; the amount of time the real-time voice signal takes to travel from the talker's mouthpiece to the listener's earpiece (also known as *mouth-to-ear* or *M2E delay*).

**Delay Jitter**    This refers to the variation of packet inter-arrival time to a destination station or terminal equipment.

**DHCP**    Dynamic host configuration protocol; an IETF protocol (RFC 2131) for passing client configuration information to hosts in a TCP/IP network.

**DiffServ**    Differentiated services; this refers to a scalable IETF protocol (see e.g., RFC 2474, RFC 2475, and RFC 2638) which performs classification of packets into a small number of aggregated flows or classes using the DiffServ codepoint (DSCP) in the IP header, and at each DiffServ router, the packets are routed on the basis of "per-hop behavior" (PHB) invoked by the DSCP. Assured forwarding (AF, RFC 2597) and expedited forwarding (EF, RFC 3246 and RFC 3247) techniques have been proposed to implement mechanisms to support the quality of service requirements for loss- and delay-jitter-sensitive applications.

**DMI**    Desktop management interface; this refers to a set of standards developed by the distributed management task force (DMTF) Inc., for managing and tracking software and hardware components in a desktop device like PC, notebook computer, server, etc.

**DOCSIS**    Data over cable service interface specifications; this refers to the interface requirements for broadband data distribution services over cable TV networks using cable modem (CM) and multimedia terminal adapter (MTA) at customers' remises and cable modem termination system (CMTS) at the head-end. DOCSIS 1.1 supports end-to-end quality of service, security, authentication and accounting, so that VoIP can be delivered over cable TV networks (see e.g., www.cablemodem.com, 2001).

**DNS**    Domain name system; IETF's host computer naming convention (e.g., RFCs 1034–1035, 1591, 2136, 2181, 2535, 2929) in which the naming data are hierarchically structured into classes and zones and can be maintained independently.

**DPC**    Destination point code; this refers to the point code (PC) based address (3 bytes, in ANSI SS7) of the node (STP or SSP or SCP) to which an SS7 signaling message is being sent.

**DS**    Differentiated service, see DiffServ, as defined earlier.

**DSL**    Digital subscriber line; this refers to a set of technologies—for example asymmetric DSL or ADSL, symmetric DSL or SDSL, high-speed DSL or HDSL, very high-speed DSL or VHDSL, etc.—that use the upper frequency band (20 KHz to ~140 KHz for upstream signal from home or office, and ~140 KHz to 1100 KHz for downstream signal to home or office) in twisted-pair copper telephone line for simultaneous transmission of multiple voice conversations and high-bit-rate data services (detailed information on DSL can be found at www.dslforum.org, www.dsllife.com, www.dslreports. com, 2001, etc.).

**DSLA**   Digital speech level analyzer; a tool for predicting speech quality and measuring the characteristics of the speech channel (see www.malden. co.uk/products/dsla/dsla.htm for details).

**DSLAM**   Digital subscriber line access multiplexer; this refers to a network element residing in the PSTN central office (CO) which multiplexes (or combines) signals from multiple DSL customers, and splits the information so that voice call related traffic can be routed to the PSTN switch, and data traffic can be routed to the Internet backbone.

**DSL Forum**   This refers to a forum of computing and telecommunication equipment manufacturers and service providers, which facilitates development of specifications (available at www.dslforum.org/aboutdsl/tr_table. html) for configuring, provisioning, and interoperability of DSL-based network elements in order to promote the DSL technology to the residential and business customers.

**DSP**   Digital signal processor or processing; processor refers to special purpose integrated circuit chips for computationally intensive processing— coding/decoding, modulation/demodulation, echo and noise cancellation, tone detection, etc.—of voice or video signal; Processing refers to algorithm-based operation of analog information which has been converted into a digital format.

**DTMF**   Dual tone multifrequency; representation of each digit (0 to 9) and characters (*, #, A–Z) using a pair of sine waves chosen from eight (four from 697 to 941 Hz and four from 1209 to 1633 Hz) different frequencies; for example, the digit 0 is represented as the combination of 941-Hz and 1336-Hz signals.

**E&M**   Ear and mouth or receive and transmit; the signaling technique that is normally used on trunks between PBX types of equipment.

**EC**   Echo cancellation; the process of removing echo from the line by keeping a sample of the speech sent on the forward path and subtracting it from the inverse of the speech coming back from the reverse direction (echoes are usually caused by a mismatch in impedance in the telephone wiring).

**EFM**   Ethernet in the first mile; this refers to an Industry alliance to develop technologies to support transmission of Ethernet frames directly over e.g., DSL removing the need to use the ATM in layer 2 (or link layer). Point-to-point connection over single-pair of voice-grade twisted-pair copper wire, and point-to-point and –multipoint connections over optical fiber links will be supported. EFM is scheduled to be lab- and field-tested during 2003, with a plan for endorsement by the IEEE P802.3ah committee in late 2003 (www.ieee802.org/3/efm, www.efmalliance.org, 2001).

**ENM**  Enterprise network management; a collection of tools and systems that is utilized to manage a network that facilitates enterprise wide computing and communications.

**Enterprise Network**  A network that facilitates voice, data, and video communications within the logical boundaries of an enterprise or corporation. Traditionally, multiple physical networks—for example, one based on PSTN and the other based on X.25, IP, frame relay (FR), ATM, and so on—are used. However, these networks are converging toward the use of a single IP-based network.

**ENUM**  Electronic numbering; IETF's approach (RFCs 2806, 2916, 3026, etc.) for mapping telephone numbers, that is E.164 addresses, into uniform resource identifiers (URIs, RFC 2396), URLs or e-mail addresses, and vice versa.

**ERP**  Enterprise resource planning; a system for managing the operations and planning the growth of all of the assets (software, hardware, network, business process, inventory, finance, etc.) with an Enterprise.

**ETE**  End-to-end; this is utilized to characterize a parameter—for example delay—from the point of origination (or source of traffic) to the destination (or traffic sink).

**Feature Server**  A server which hosts various telephone call related features, and interacts with PSTN's IN/AIN hosts to deliver call features to the customers.

**FCC**  Federal Communications Commission; this refers to an independent Government agency (www.fcc.gov) of the USA, which regulates local, long-distance, and International (telephone and information) communications.

**FEC**  Forward error correction; a mechanism that calls for addition of extra bits—generated by using a structured algorithm such as Reed Solomon coding—to a packet; these extra bits can be used to reconstruct the information in the original bit stream in case of error in or loss of information. For voice transmission using IP, the IETF has recommended various options (RFC 2354) for packet repair using FEC.

**Firewall**  Software-and/or hardware-based pinhole opening and closing mechanisms to allow authorized and traceable access to a private or internal packet-based network.

**FR**  Frame relay; a connection-oriented link or layer-2 (of the OSI model) protocol, which support a maximum of 4096 Bytes of frame.

**Framing**  Encapsulation of a segment of packetized information (data, speech or voice sample, video, etc.) using a header and trailer. The header contains addressing and routing information, and the trailer contains error detection and correction codes.

**FTP**   File transfer protocol; an IETF protocol (RFC 959) that is an Internet application commonly used to transfer files in a TCP/IP network. The TFTP is a trivial version of the FTP.

**FXO**   Foreign exchange office; this refers to an interface on a VoIP device that mimics a standard telephone handset, i.e., it requires another device to provide it a dial tone. A VoIP device with FXO can be connected to an analog PBX extension jack.

**FXS**   Foreign exchange station; this refers to an interface that mimics the public switched telephone network (PSTN), i.e., it provides dial tone to a standard telephone handset. A VoIP device with FXS can be connected directly to a phone, fax, central office port, PBX, key telephone system, etc.

**G.114**   An ITU-T recommendation specifying that for toll-quality voice, the maximum allowable one-way (talker's *mouth to* listener's *ear*) delay should not exceed 150 msec.

**G.168**   An ITU-T recommendation that specifies electrical line echo cancellers—by subtracting an estimated echo from the circuit echo—when the echoes are caused by two- to four-wire conversion hybrids. Echo cancellers are voice-operated devices placed in the four-wire portion of the circuit to improve voice quality (a 128-msec echo canceller tail is needed for carrier class or toll-quality voice).

**G.711, G.723, G.729**   These are ITU-T standards for speech coding for real-time voice communications; G.711 uses pulse code modulation (PCM) scheme, and generates a bit stream of 64 Kbps of speed, G.723 uses multipulse maximum likelihood quantization (MP-MLQ) technique to generate a bit stream of 6.3 Kbps of speed or algebraic-code-excited linear-prediction (ACELP) technique to generate 5.3 Kbps bit stream, and G.729 uses conjugate structure algebraic code excited linear prediction coding (CS-ACELP; this refers to algorithmic compression of digitized speech using human vocal tract model), and generates a bit stream of 8 Kbps of speed.

**GK**   Gatekeeper; ITU-T's H.323 element (router or server) that maintains registry of GW and terminal equipment devices in a multimedia network. It controls access to LANs and provides address translation, connection control and routing, bandwidth management, finding GWs, and other services to the H.323 terminals and GWs.

**GKAPI**   Gatekeeper application programming interface; an API that can be used to facilitate communications of the applications with the GK.

**GKTMP**   Gatekeeper transaction message protocol; this is the protocol that is used for communication with back end non-Cisco-IOS (Internet operating system; see www.cisco.com) servers.

**GMPLS**   Generalized multiprotocol label switching; a generalized version of the MPLS protocol (*see MPLS*; this can be found at www.mplsforum. org), which includes signaling for delivering QoS in IP-based optical networks.

**GPS**  Global positioning system; this refers to the most authentic system for capturing and distributing precise time and time interval (details can be found at http://tycho.usno.navy.mil/gps.html).

**GR**  Generic requirements; the documents, which are prepared and published by Telcordia (www.saic.com/about/companies/telcordia.html, 2001) to specify the Telecom network, switching equipment, and service requirements.

**GW**  Gateway; a network element that repackages TDM-formatted speech or voice signal from a circuit-switched call into RTP/UDP/IP packets and/or AAL-x/ATM cells. In the context of ITU-T's H.323 recommendation, a GW is an element that provides real-time two-way communications between H.323 terminals on the LAN and other ITU-T terminals in the WAN or to another H.323 GW.

**HFC**  Hybrid fiber coax; a network where the access or distribution system utilizes coaxial cable, and the backbone or transport network uses fiber optic transmission system. Cable TV service providers commonly use HFC networks.

**H.323**  An ITU-T specification for real-time multimedia communications over LANs where the QoS cannot be guaranteed.

**HTTP**  Hypertext transfer protocol; this refers to a stateless request/response-type IETF protocol (see e.g., RFC 2616, RFC 1945) that is utilized for formatting and transmitting messages from Web browser to Web server, and for receiving files from Web server to the Web browser.

**HVB**  Hammer visual basic; a visual basic language developed by Hammer (now a part of Empirix, www.empirix.com, 2002) for scripting IP telephony tests and measurements programs or suites.

**IA**  Implementation agreement; the documents prepared by the multi-service switching forum (MSF, www.msforum.org, 2002) to specify requirements of the interface between the components of a multi-service switching system in order to guarantee interoperability.

**IAM**  Initial address message; an ISUP message to initiate a telephone call setup and routing using the SS7 network.

**IAD**  Integrated access device; this refers to a customer premise device which supports voice and data communications services using PSTN and IP domain signaling, call control and access methods over only one set of wires or connection to the access network. For example, circuit-switched, SIP, H.323 etc. based call control and Ethernet (IEEE 802.3) based LAN access, etc. are commonly supported through an IAD (can include DSL modem or Cable modem) to the customers.

**ICI**  Interexchange carrier interconnection; one or more procedures for connecting calls between dissimilar networks involving an IXC, irrespective of whether the call stays within a LATA or crosses the LATA boundary (see, e.g., Telcordia's GR-394-CORE).

**IETF**   Internet engineering task force; the organization that issues requests for comments (RFCs) to develop open protocols for the Internet (details can be found at www.ietf.org).

**ILEC**   Incumbent local exchange carrier; a local communication (primarily access) service provider that offers services in a geographical area using its own network infrastructures (lines, switches, routers, servers, computers, etc.).

**IMT**   Inter-machine trunk; this refers to high-capacity (T1, T3, etc. in North America) trunk between two SS7-controlled PSTN switches. The endpoints of an IMT are identified by OPC (origin point code) and DPC (destination point code), and the channels within the IMT trunks are usually controlled by circuit identification code (CIC).

**Intranet**   This refers to corporate or Enterprise networks that facilitate seamless communications and networked computing within a single Corporation. Most of today's Intranets use the Internet protocol (IP) based networking, although other technologies like X.25, Frame relay (FR), Asynchronous transfer mode (ATM), etc are also utilized.

**IntServ**   Integrated services (Intserv); this refers to an IETF architecture (see e.g., RFC 2998, RFC 1633) which supports a mechanism (e.g., RSVP based signaling) for delivering end-to-end quality of service (QoS) to applications running over heterogeneous networks.

**IP**   Intelligent peripheral in the context of AIN, and refers to a network element which hosts PSTN call or connection resources which are needed for conferencing, speech synthesis, IVR, etc. IP also means Internet Protocol in context of the Internet, and refers to an IETF protocol (IP version 4 is defined in RFC 791, and IP version 6 is defined in RFC 2460 and RFC 1883) which operates at layer-3 (network layer of the OSI model) for connectionless and best-effort/unreliable internetworking of heterogeneous networks.

**IP Centrex**   Internet protocol based centrex; this refers to the delivery of PSTN domain centrex features and services using an Internet Protocol (IP) based private network and network elements like IP-PSTN media gateway, centrex feature gateway, call controllers, etc.

**IPDC**   Internet protocol device control; an IP appliance or device control protocol developed by a Level-3-Led consortium of vendors and service providers; IPDC has been merged with SGCP to develop MGCP.

**IP-PBX**   Internet protocol based PBX; this refers to a system that is capable of providing traditional circuit-switch and packet-based PBX functions using the Internet protocol (IP) based software and hardware/network-element.

**IPSec**   Internet protocol (IP) security protocol; a suite of protocols that can be used to secure communication at the IP (layer-3 or network layer in the OSI model) between two peers. The suite consists of IP security architecture (RFC 2401), key management protocol like the Internet key exchange (RFC 2409), and traffic protection protocols like the IP authentication header (RFC 2402) and the IP encapsulating security payload (RFC 2406).

**IP Telephony**  This commonly refers to delivering basic and advanced telephony services using IP based devices (e.g., IP phones), network (e.g., the Internet), and network elements (e.g., VoIP call server, IP-PSTN media gateways, etc.) for call control, signaling, and voice transmission. Very often, IP Telephony and VoIP are used synonymously.

**ISC**  International softswitch consortium, this refers to an Industry forum (www.softswitch.org) that intends to promote openness and Interoperability of Internet based real-time multimedia communications.

**ISO**  International standardization organization; a multi-national group that develops standards on a variety of topics (see, www.iso.org). ISO created the well-known OSI model for structured communication in a packet-switched network environment.

**ISDN**  Integrated services digital network; an ITU-T/ANSI standard for delivering signaling (D channel, which is 16 Kbps in BRI and 64 Kbps in PRI) and media transmission (2B to 23B, where B refers to a 64 Kbps channel) services over separate logical channels using the same physical link to the customer's premises; both basic rate interface (BRI or 2B+D) and primary rate interface (PRI or 23B+D) have been defined for narrowband (N-ISDN) services, whereas broadband ISDN (B-ISDN) was expected to use the ATM/SONET technology.

**ISP**  Internet service provider; this refers to a type of company who provides access to the public internet for basic e-mail, web-based, file transfer, etc. services over dial-up connection using twisted pair copper wire based telephone lines; enhanced ISPs offer advanced data services to high-end residential customers and to Enterprise customers using DSL line, Cable TV channel, etc.

**ISUP**  ISDN user part; messages and a protocol in the SS7 network that define the parameters and events used to set up and tear down telephone calls in PSTNs.

**ITU-T**  International telecommunication union—telecommunication standardization sector; a subdivision of the ITU (www.itu.int) originally known as the international telegraph and telephone consultative committee (CCITT). The purpose of ITU-T is to develop telecom networking and service–related standards with participation of service providers and vendors from all over the world.

**IVR**  Interactive voice response; a system that allows a caller to retrieve and listen to prerecorded voice files using DTMF tone(s) or prespecified voice tone(s) as triggers or inputs.

**IXC**  Interexchange carrier; a carrier that interconnects the local exchange or central office or CO switches of LATAs in different geographical locations.

Traditionally, IXCs are the long-distance voice communication service providers such as AT&T, MCI-WorldCom, and Sprint.

**JAIN**    Java API for intelligent networking; this refers to a set of Sun Microsytem's Java programming language based APIs for rapid development of portable Next-generation services which allow converged and secure access to services from any—telephony and data—network (details can be found at java.sun.com/products/jain/). A similar API is Parlay (described later).

**JTAPI**    Java telephony API; this refers to Sun Microsystem's Java programming language based API for development of applications which can invoke telephony services like dialing out a number, responding to a phone call, collecting information from the caller by using IVR system, etc.

**LAN/MAN/WAN**    Local area network/metropolitan area network/wide area network; these refer to packet based network infrastructures with different size of geographical area coverage, e.g., a LAN can provide data networking over a building or a campus using Ethernet and TCP/IP networking, a MAN can provide data networking over a Metropolis using Frame Relay (FR), ATM, IP, and Gigabit Ethernet (GigE) networking, and a WAN can provide data networking over the whole country or continent using a variety (IP, T1, FR, SONET, etc.) of networking technologies.

**LATA**    Local access and transport area; a geographical area within which a regional Bell operating company (RBOC) is allowed to offer exchange telecommunications and exchange access services.

**Latency**    This refers to the amount of time lapsed between a request and the corresponding response, e.g., the access latency is defined as the amount of time between the time instant when a device requests for access to a network and the time instant when it actually is granted permission to transmit. In the context of VoIP, latency refers to the amount of time delay suffered by a voice packet when it is transported over an IP network, and ideally this delay value should be the same (i.e., a constant value) from source to destination for all of the successive voice packets for a VoIP session or call.

**LD**    Long distance; this refers to long-distance (or inter-LATA) voice communication service; the two LATAs may be located in the same State (e.g., within Massachusetts, USA) or in two different States (e.g., one in New York, USA, and the other in Texas, USA). Note that the data services are typically all-distance type service, and are not charged on the basis of the distance traveled by the signal.

**LDAP**    Lightweight directory access protocol; this refers to a TCP/IP based lighter (than X.500 directory access protocol) information directory access protocol (IETF's RFC 1777) for on-line services like e-mail address, phone number, etc. lookup.

**LDP**    Label distribution protocol; a protocol or a set of procedures which the label switched routers (LSRs) use to inform the label or forwarding equivalence class (FEC) of packets to each other (see IETF's RFC 3031 for further details).

**LEC**  Local exchange carrier; a carrier—for example, any of the independent LECs and regional Bell operating companies (RBOCs)—that provides traditional voice telephony services in a LATA using local exchange or central office or CO switches (e.g., Telcordia's GR-2982-CORE).

**LMDS**  Local multi-point distribution service; a wireless local loop (WLL) system that operates at 27.5–29.5 GHz band, and can provide service to customers within 5 kilometer of radius.

**LNP**  Local number portability; portability of a local telephone number when a customer moves from one locality (server by one central office or CO switch) to another (server by another CO switch). In the new locality, a local routing number (LRN) is assigned to the customer's telephone number in the local service management system (LSMS) database for use by the LECs, IXCs, and IECs for call routing. The Federal Communications Commission (FCC) has appointed the Lockheed Martin Company to manage the Number Portability Administration Center (NPAC), which coordinates LNP and LRN in the LSMS databases.

**LSP**  Label switched path; the path defined by a sequence of label switched routers (LSRs)—at one level of the hierarchy—that is followed by a packet in a particular forwarding equivalence class (FEC, see IETF's RFC 3031 for further details).

**LSSGR**  LATA switching systems generic requirements; this is a collection of Telcordia's 120+ GRs which describe generic characteristics and utilities required for traditional PSTN-based voice call processing and signaling including CLASS, features, and vertical service capabilities required to offer local and intra-LATA services, and to provide access to Domestic/National (inter-LATA) and International carriers.

**L2TP**  Layer two tunneling protocol; this refers to an IETF protocol (RFC 2661, RFC 2809, RFC 2888, RFC 3070, RFC 3193) that defines an encapsulation mechanism for multiplexing multiple, tunneled PPP sessions for creating tunnels between two nodes. This method essentially extends the PPP model to allow the layer 2 (L2) and PPP endpoints to reside on different packet-switched network devices.

**MAC**  Media access control; the protocol which controls access to and from a computer system via a NIC to a network. MAC functions reside in the lower sub-layer of the data link layer (second layer) of the OSI model.

**MAN**  Metropolitan area network, see LAN/MAN/WAN for details.

**MC**  Multipoint controller; an ITU-T H.323 entity on a LAN that controls three or more terminals participating in a multipoint conference. It may also control conference resources such as who is multicasting video, but it does not perform mixing or switching of audio, video, and data.

**MCU**  Multipoint control unit; this refers to an ITU-T's H.323 endpoint on a LAN that provides the capability for three or more terminals and gateways to participate in a multipoint conference; the MCU may consist of two parts: a mandatory multipoint controller (MC) and an optional multipoint processors (MPs); in the simplest case, an MCU may consist of only an MC, with no MPs.

**Megaco** Media gateway control; an ITU-T and IETF protocol that can be used between a physically decomposed MGW and an MGC. In ITU-T it is known as the *H.248 recommendation*, and in IETF it is called the *RFC 3015*.

**MG** or **MGW** Media gateway; the functions of converting media or bearer information from one format (e.g., TDM format in a PSTN network) into another (e.g., RTP streams in an IP network, AAL2 and/or AAL5 cells in an ATM network) and vice versa.

**MGC** Media gateway controller; this refers to the function of call state control for the media channel or connection which is being supported by a media gateway. In H.323 protocol, the gatekeeper (**GK**) performs this function, and in case of MGCP the call agent (**CA**) performs this function. Sometimes, MGC is also referred to as VoIP call controller or call manager or VoIP call server.

**MGCP** Media gateway control protocol; a combination of the simple gateway control protocol (SGCP) developed by Cisco and Telcordia (formerly Bellcore) and the Internet protocol device control (IPDC) protocol developed by a Level-3-Led consortium of vendors and service providers. MGCP facilitates physical separation of call control and management from the MGWs that provide service at the edge of the networks so that call control software can run on a general-purpose computing platform to provide call and MGW control functions. It is also an IETF recommended protocol (RFC 2705 and RFC 2805) for providing voice services over IP.

**MGTS** Message generator traffic simulator; a test equipment (originally from Tekelec, www.tekelec.com, but now from Catapult, www.catapult.com/ mgts/mgtsover.htm, 2002) which can be utilized to emulate SCP, SSP, and STP functions of the SS7 network.

**MIB** Management information base; a collection of managed objects stored in a database. These objects maintain information related to network elements' configuration, status, performance, etc. The database can be monitored by a network management system like SNMP (defined later).

**MIME** Multi-purpose internet mail extensions; this refers to an IETF specification (see e.g., RFC 1522, RFC 1521, and RFC 1341) which defines the mechanisms for formatting non-text (audio, video, graphics, etc) messages for transmission over the Internet.

**MMDS** Multi-channel multi-point distribution service; a wireless local loop (WLL) system that operates at 2.15–2.70 GHz band, and can provide service to customers within 50 Kilometer of radius.

**MOS** Mean opinion score; this refers to ITU-T's P.800 recommendation for subjective measurement of speech transmission quality. MOS is calculated by taking weighted average of voice quality scores ("1" means "bad,"

"2" means "poor," "3" means "fair," "4" means "good," and "5" means "excellent") assessed by a group of men and women of different race, culture, and tone/accent. Usually, a set of pre-selected medium/long voice samples are played over the same transmission media under different types of impairment conditions, and the group of men and women listen to the samples at the other end of the channel to assign a MOS number for speech quality ("1" for bad quality, and "5" for excellent quality). Note that the "toll-quality" voice has a MOS of 4.

**MP** Multipoint processor; an ITU-T H.323 entity on a LAN that provides (a) centralized processing of audio, video, or data streams in a multipoint conference and (b) mixing, switching, or other processing of media streams under the control of the MC. It may also process a single media stream or multiple media streams, depending on the type of conference supported.

**MPLS** Multiprotocol label switching; an IETF protocol (see, e.g., RFC 2702 and RFC 3031/2/5/6; also see www.mplsforum.org and www.mplsrc.com for details) that enables delivery of a different class/category and QoS over the traditional IP-based network by using special "labels" in the packet header.

**MP-MLQ** Multi-pulse maximum likelihood quantization; a technique utilized by G.723 voice coding scheme to generate a bit stream of 6.3 Kbps of speed.

**MSF** Multi-service switching forum; this refers to an international association of telecommunications and data communications service providers and equipment manufacturers with a mission to develop open protocol based systems for multi-service switching (the implementation agreements are available at www.msforum.org/techinfo/approved.shtml).

**MSS** Multiservice switch; a network element that can support a variety of protocols and interfaces for interworking between PSTN and packet networks.

**MTA** Multimedia terminal adapter; the multimedia information adapter—within or adjacent to the cable modem in customer's premises—of the DOCSIS (defined earlier) standard.

**M2E** Mouth to ear; an identifier which is commonly used to refer to the voice signal transmission delay from talker's (e.g., the calling party's) mouth to listener's (e.g., the called party's) ear in a telephone call.

**MTP** Message transfer part; defines functions for reliable transfer of SS7 messages from one signaling point to another in a fashion which is equivalent to OSI layer-1 to layer3; defined in ITU-T documents Q.700 through Q.709 and ANSI Standards T1.111-1992.

**MTU** Maximum transmission unit; the maximum packet size, in bytes, that a particular interface supports (receives, processes, and transmits).

**MU-Law** or **$\mu$-Law** An ITU-T specification for logarithmic conversion between analog and digital signals for pulse code modulation (PCM) technique in G.711 coding; used mainly in North America and Japan (A-Law is used in Europe; *see A-Law*).

**NAT**    Network address translator; this refers to a table-driven mechanism (RFC 1631) for reusing IP addresses by assigning multiple local IP address to one global IP address (used for classless Inter-Domain routing).

**NAPTR**    Naming authority pointer; this refers to a protocol (IETF's RFC 2915) for obtaining universal resource information (URI) or media-specific end-point—e.g., a wire-line phone, fax machine, wireless phone, e-mail address, etc.—contact information record by using a DNS query.

**NE**    Network element; a computer server or switching equipment that facilitates network based communication services.

**NEBS**    Network equipment building system; a set of Telcordia (formerly Bellcore) generic requirements (GRs) for building a disaster-proof (fire, flood, earthquake, etc.) enclosed space for housing PSTN CO equipment (see, e.g., GR-63-CORE, GR-2930-CORE, etc.).

**NGN**    Next generation network; this refers to a packet based distributed network and system architecture for delivering feature-rich voice, data, and video (i.e., integrated) services on demand using mostly the Internet technologies.

**NGEN**    Next generation enterprise network or networking; this refers to a packet-based unified or integrated IP-based network infrastructure for Enterprise or Corporation wide computing, and voice, data, and video communications.

**NIC**    Network interface card; an adapter or a card which facilitates communication from a computer to a network. For example, an Ethernet NIC allows access to an Ethernet LAN for data transmission and reception using the Ethernet-based media access control (MAC) protocol.

**NIST**    National Institute of Standards and Technology. The advanced networking technology division of NIST has developed an IP network emulation tool called NIST-Net that can be found at www.antd.nist.gov/nistnet/, and the Internet Telephony/VoIP group is developing tools for speech quality measurement (details are available at www.antd.nist.gov).

**NNI**    Network to network or node interface; the interface between two network (private or public) level ATM switches, as defined by the ATM Forum.

**NTP**    Network time protocol; this refers to an IETF protocol (RFC 1305) which runs using UDP/IP to help adjust client's local clock with network time server's or NTS's clock (in client-server environment), the NTS usually derives clock from a coordinated universal time (UTC), like GPS; the timing accuracy in NTP varies from fraction of a msec in LAN to tens of msec in WAN.

**NTS**    Network time server; this refers to a server which provides timing information (clock) to IP domain network elements using the NTP (described above). NTS may be configured and instrumented to derive clock from a GPS receiver.

**OPC** Origin point code; this refers to the point code (PC) based address (3 bytes) of a node from which an SS7 signaling message originated.

**OSI** Open system interconnection; a seven-layer functional model developed by the International Standardization Organization (ISO) to facilitate structured communication in a packet-switched network environment; the seven layers are physical (layer-1), Data-Link (layer-2), Network (layer-3), Transport (layer-4), Session (layer-5), Presentation (layer-6), and Application (layer-7).

**Packet Loss** Dropping of packets in a packet-switched network due either to corruption of the packet header or to buffer overflow in the routers. Loss of packets causes degradation of voice quality.

**PAMS** Perceptual analysis/measurement system; a voice band (300 Hz to 3400 Hz) speech coding assessment method which uses auditory model to compare original and transmitted (degraded) voice signals (see www. malden.co.uk for further details).

**Parlay** A set of open APIs—consisting of framework and services interfaces—developed by the Paraly group (see www.parlay.org for details) that can be used to link information technology applications (such as IP messaging, control, security, and performance management) to telecom network–based services such as call control. A similar API is JAIN (*see JAIN*).

**PBX** Private (automatic) branch exchange; a CLASS-6 PSTN switch that resides in the customer's premises, provides plain/advanced telephony services to at least 20 customers, and supports connectivity to a CLASS-5 PSTN switch and/or a private network.

**PC** Personal computer in the context of the computer industry and point code in the context of telecommunications network operations. Point codes are decimal digit–based addresses of SSPs, SCPs and STPs, and these addresses are used for message routing in an SS7 network.

**PCM** Pulse code modulation; a technique for digitizing analog signal, e.g., voice signal, by sampling of the analog waveform periodically, and converting the samples into digital codes. For telephony applications, voice signal is sampled 8000 times per second, and 8 bits are used to code each sample, and this produces a stream of 64,000 bits/sec of data.

**PESQ** Perceptual evaluation of speech quality; a voice band (300 Hz to 3400 Hz) speech coding assessment method which uses sensory model to compare original and transmitted (degraded) signals. PESQ was developed by combining PAMS and PSQM+, and the ITU-T's study group 12 (SG-12) has recently approved it as its P.862 recommendation (see www.pesq.org for further details).

**PIN** Personal identification number; a password that can be used with an account number for authentication purposes.

**PINT**    PSTN/internet interworking; this refers to an IETF activity (see e.g., IETF's RFC 2458, RFC 2848, RFC 3055) for developing IP based protocols and standards for making PSTN service invoke-able by the Internet clients (PC, IP phone, etc.).

**POP**    Point of presence; an access point which is commonly used by the data communications service providers (e.g., an Internet service provider or ISP) to aggregate customers' access to the network. POP also stands for post office protocol (IETF's RFC 1939) which is used to retrieve email from a mail server.

**POTS**    Plain old telephone service; the basic (voice-only) telephone service that a PSTN delivers using a standard single twisted-pair copper wire telephone line or a DS0 line.

**PPP**    Point to point protocol; this refers to an IETF protocol (RFC 1547, RFC 1661) that defines encapsulation, link control, and network control mechanisms for transmission of multi-protocol data-grams over point-to-point links.

**PPTP**    Point to point tunneling protocol; this refers to an IETF protocol (RFC 2637) that defines mechanisms for carrying multiple PPP data-grams between PPTP access concentrator (PAC) and PPTP network server (PNS) over a tunnel, with the control channel running over a TCP connection.

**PRI**    Primary rate interface; the ISDN PRI interface consists of 23 B channels (each 64 Kbps) and one data or signaling channel of 64 Kbps. Thus, one PRI link becomes 1536 Kbps channel.

**PSAP**    Public safety answering point; this refers to a PSTN-hosted call center where the emergency calls (e.g., the 911 calls in the USA) are routed. The operator dispatches a 911 call with information on caller's physical location to the local fire, police, and medical emergency response teams.

**PSQM**    Perceptual speech quality measurement; this refers to an ITU-T recommended technique (P.861 recommendation) for objective assessment of voice band (300 Hz to 3400 Hz) speech codecs. PSQM repetitively uses fast Fourier transform (FFT), normalization, and sliding/overlapping computation windows to compare original and transmitted (degraded) signals; a PSQM score of zero means best quality of transmission, and 6.5 means worst transmission quality. PSQM+ is a supplement to PSQM, and it takes packet loss and difference in perception due to sound volume into consideration for assessing speech quality (see www.psqm.com for further details).

**PSTN**    Public switched telephone network; usually the local telephone networks maintained by the regional Bell operating companies (RBOCs).

**PUC**    Public utility commission; a local or statewide commission that administers consumer-interest-focused regulations for services of basic utilities such as telephone, water, and electricity.

**Q.931**    An ITU-T specification for a message-based (layer-3) out-of-band signaling protocol for call control between the user and the user network interface (UNI).

**QoS**  Quality of service; the level of service—in term of transmission delay, delay jittter, packet loss, and so on—required for a specific application such as VoIP (a related term is CoS; *see Cos*).

**RADIUS**  Remote access dial-in user service; this is a database service which is used to authenticate modem and ISDN connections, and for recording session or connection hold time. RADIUS servers can be utilized to offer AAA (defined earlier) services.

**RAS**  Registration, admission/administration, status; this is a part (H.225) of the ITU-T H.323 protocol which allows communication between an H.323 gatekeeper and a gateway. RAS also stands for remote access service.

**RED**  Random early detection; a TCP/IP congestion management mechanism for controlling the queue size in a router by either drooping the packets (once a pre-set threshold is exceeded) or marking them as discard-eligible. This method was originally proposed by S. Floyd and V. Jacobson in "Random Early Detection gateways for Congestion Avoidance," IEEE/ ACM Transactions on Networking, Vol. 1, No. 4, pp. 397–413, August 1993.

**Regulatory Features**  The voice telephony features that a company must support if it wants to be a licensed carrier; the features include 911 (emergency service) and 411 (directory assistance) calling, CALEA (communications assistance for law enforcement agencies), TRS (telecom relay service), and others.

**REL**  Release message; an ISUP message for releasing a telephone connection.

**RFC**  Request for comment; RFCs are IETF-controlled documents, which are utilized for open discussion before standardizing protocols and services for the Internet.

**RGW**  Residential gateway; a gateway that provides an interworking function between one or more analog lines and a packet network within a residence or the customer's premises.

**RLC**  Release complete message; an ISUP message for completing release of the circuits which have been or are being used for a telephone call.

**RSVP**  Resource reservation protocol; the IETF protocols (RFC 2205 to 2210 and others) that define the ability to dynamically reserve or allocate bandwidth and latency to a particular traffic flow in an IP network.

**RTCP**  RTP control protocol; a mechanism for controlling the RTP session by periodically transmitting control packets to all of the participants of a session using the same mechanism which is used for data packets (see IETF's RFC 1889 for details).

**RTP**  Real-time transport protocol; an IETF protocol (RFC 1889) for real-time transmission of streaming media (e.g., real-time VoIP). It is a part of the ITU-T's H.323 specification for real-time multimedia communications over LANs.

**SAD** or **VAD**    Speech or voice activity detection; detecting the talk spurts and silence intervals during a real-time telephone conversation with the objective of packetizing and transmitting the talk spurts only and of using comfort noise generation (CNG) at the receiver instead of transmitting silence.

**SAN**    Storage area network or networking; this refers to a very high-speed network of storage devices (disk, tape, optical, etc.) associated with data servers over a very small network, e.g., within a 10 ft. × 10 ft. × 10 ft. room which can support physically remote backup and archive facilities by maintaining mirror image of locally stored information in remote storage devices. SANs are typically used in medium and large Enterprises for Web hosting, networked computing, etc. applications (see www.snia.org, 2001 for details).

**SAP**    Session announcement protocol in the context of SIP (defined later), and Service Access Point in the context of ISO's OSI model (defined earlier).

**SCP**    Signal or service control point; a host computer or server that maintains service logic for AIN/IN, the database of the telephone numbers, and their mapping to one or more phone numbers for LNP, 8xx, and 10-10-xxx call processing and routing.

**SCTP**    Stream control transmission protocol; an IETF protocol (RFC 2960, RFC 3057) that provides better security, timing, and reliability than the existing TCP/UDP-based transport mechanism. The primary features of SCTP are (a) acknowledged, error-free, nonduplicated transfer of user data, (b) data segmentation to conform to the discovered path message transmission unit (MTU) size, (c) in-sequence delivery of user messages within multiple streams, (d) optional multiplexing of user messages into SCTP datagrams, (e) network-level fault tolerance through support of multihoming, and (f) backward compatibility with UDP. SCTP addresses the transport of SS7 signaling messages such as ISDN (Q.931), ISUP, and so on between various network elements—such as the SG, MGC, and MGW—over IP-based networks.

**SDP**    Session description protocol; this refers to an ASCII text based IETF protocol (RFC 2327) which is used in SIP to describe the features, length, recipient(s), etc. of multimedia streams in a session.

**SG** or **SGW**    Signaling gateway or SS7 gateway; this refers to the functions of translating, terminating, and relaying of PSTN native signaling (SS7) messages to and from the edge of a data (mostly IP-based) network. **SG** is also utilized to refer to "study group" when it is used in conjunction with the activities of a focused group within a Standardization committee.

**SGCP**    Simple gateway control protocol; an MGC protocol developed jointly by Cisco and Telcordia (formerly Bellcore); SGCP has been merged with IPDC to develop MGCP.

**SIP**    Session initiation protocol; it is an IETF protocol (RFC 3261, RFC

3262, RFC 2543) for telephony and multimedia call control and signaling over the Internet, using *mostly* the Internet paradigm—that is open, distributed, scalable, low-cost, etc.—and protocol (details are available at www.ietf.org/html.charters/sip-charter.html, www.sipforum.org, and www.cs.columbia.edu/~hgs/sip).

**SIP-PS** Session initiation protocol-proxy server; a surrogate server that mediates SIP call setup and teardown and remains in the call path for the duration of the call or session (useful for sophisticated routing and services, since the location server functionality can reside in this server as well).

**SIP-RS** Session initiation protocol-registration server and redirect server; SIP registration server that maintains records of registered users, their profiles, and the call details (AAA format); the SIP redirect server conveys the call control to the agent or device from which the call originated (i.e., the INVITE message was initiated) and does not stay in the call path after that event.

**SIP-T** Session initiation protocol-tunneling or telephony; this refers to the Session Initiation Protocol-Best Current Practice (BCP) which provides encapsulation of ISUP message with the header carrying translated ISUP routing information, and thereby enables SIP to be used for ISUP-based call setup between PSTN network and IP telephony (SIP-based) networks (BCP guidelines are defined in IETF's RFC 2026) (see also www.sipforum.org, www.sipcenter.com, etc. for details).

**SIP-UA** Session initiation protocol-user agent; a client-side application that contains both SIP-UAC and SIP-UAS.

**SIP-UAC** Session initiation protocol-user agent client; a client-side application that can initiate a SIP request.

**SIP-UAS** Session initiation protocol-user agent server; a server-side application that communicates with the user when it receives a SIP request and then responds with an accept, reject, or redirect message.

**SLA** Service level agreement; it refers to any agreement between a customer and a service provider regarding satisfying a set of quality of service (QoS) parameters—like delay or latency, delay jitter, packet loss, availability of link bandwidth, etc. for any specific session or a physical connection, mean time to respond and repair during service outage, and so on. IETF's RFC 2475 and RFC 3198 define dynamic SLA requirements for a session.

**SMDI** Simplified message desk interface; this refers to an asynchronous serial data transmission protocol which is commonly used for integrating IP telephony system with legacy (PSTN) voice mail system over analog line or RS-232 serial interface.

**SMTP** Simple mail transfer protocol; this refer to an IETF protocol (RFC 2821) that is utilized for sending electronic mail (E-mail) message between computer servers. An E-mail client can retrieve the message by using POP (post office protocol, see e.g., RFC 2449 and RFC 1734) or IMAP (Internet message access protocol, see e.g., RFC 2971 and RFC 2683).

**SNMP**    Simple network management protocol; an IETF (SNMP v3 is specified in RFCs 2271–2275) recommended and most widely used protocol for IP device control and management. It consists of SET/GET messages to facilitate configuration and status requests and "Traps" for alarms.

**Softswitch** or **softswitch**    This refers to an architecture for evolution of the monolithic PSTN system to an organization where telephony call feature hosting and delivery services, call control, media adaptation, switching and routing, etc., functions are separable. Some researchers also refers Softswitch to a combination of media gateway controller or call controller (or call server or call agent) and SS7 signaling gateway.

**SOHO**    Small office home office; a remote office with computers and telephones connected to corporate network (Intranet) to facilitate work-at-home or telecommuting.

**SONET**    Synchronous optical network; a synchronous TDM-based American National Standards Institute (ANSI) standard for connecting optical fiber–based transmission systems. It operates at the physical layer. ATM-based broadband ISDN (BISDN) runs as a layer on top of SONET as well as on top of other technologies. The corresponding ITU-T standard is called *synchronous digital hierarchy* (SDH). ATM is frequently used from DS-3 to OC-12 (622 Mbps) speed, and SONET/SDH is generally used from OC-1/3 to OC-384 (~20 Gbps) speed.

**SPIRITS**    Services in the PSTN/IN requesting internet services; An IETF activity (see e.g., IETF RFC 3136) which is focused on making the Internet domain information like Intertnet call-forwarding/call-waiting/caller ID, etc. available to PSTN clients (e.g., a black or POTS phone).

**SSP**    Service or signal switching point; an end office or central office or CO telephone switch function that has connectivity to the SS7 network for querying external service logic or to the database to facilitate call processing from an end user.

**SS7**    Signaling system no. 7; an ITU-T/ANSI recommended common channel signaling (CCS) system. It defines the protocol stack, interface, and architecture for a highly available and reliable message (ISUP, TCAP, database query, etc.) switching system for call control, routing, and management in PSTN. The SS7 network consists of switching nodes or STPs, databases or SCPs, and connecting links (T1, V.35, etc.).

**STP**    Signal transfer point; a mated-paired (to achieve high availability and reliability) packet switching node of an SS7 network that is directly connected to the PSTN switch and the SCP (telephone numbers database). STP receives, routes, and forwards call setup, call management, and call teardown messages.

**Stratum**    In PSTN networks, this refers to the survival performance (stability) of an oscillator in the case of failure of synchronization. Typically, stratum-0 refers to the reference clock source, such as the GPS, USNO, NIST, or other clock; stratum-1 is the primary time server and has an accuracy of $1.0 \times 10^{-11}$, stratum-2 is the secondary time server and has an accuracy of $1.5 \times 10^{-8}$, stratum-3 has an accuracy of $4.5 \times 10^{-6}$, stratum-4 has an accuracy of $3.5 \times 10^{-5}$, and so on (up to stratum-15).

**SVC**   Switched virtual circuit; a virtual or emulated circuit which is established between two end-points for the duration of a session or service using a standard signaling method.

**T.4**   An ITU-T protocol that describes formatting of page image data in fax transmission.

**T.30**   ITU-T's fax session control protocol that describes the formatting of nonpage data such as capabilities negotiation messages in fax transmission.

**T.120**   A portion of the ITU-T H.323 specification related to data-sharing applications (e.g., white boarding).

**TAPI**   Telephony API; this refers to an API which enables development of computer-based applications to dial a telephone number, store commonly dialed numbers, record greetings, take dictation using speech recognition, etc.

**TCAP**   Transaction capabilities application part; SS7 messages and an application-level protocol for exchanging any transaction-related information (not call or circuit control) between two communicating application processes (e.g., Telcordia's GR-1129-CORE). TCAP is used for both database-oriented (e.g., calling card, 8xx, AIN) and switch-to-switch services including repeat dialing and call return.

**TCP**   Transmission control protocol; an IETF protocol (RFC 793) operating in layer-4 (transport layer of the OSI model) that can be used for reliable communication between host computers in a packet-switched environment. Reliability is achieved by using flow control, acknowledgment of packet reception, and sequence numbers in the packet header.

**TDM**   Time division multiplexing; a multiplexing techniques in which each user is assigned to a time slot in a round-robin fashion for accessing the channel (communication medium).

**TE**   Terminal equipment; an endpoint in a network that can be used for bidirectional real-time communications, e.g., in H.323 a TE can set up a call to another TE, GW, or MCU either directly or via a GK.

**TG** or **TGW**   Trunking gateway; a trunking-level media adaptation device that transforms TDM-formatted media (e.g., speech or voice signals) into one or more packet-based (e.g., IP, ATM) formats so that information transmission can occur over a packet network.

**TMN**   Telecommunications management network; this refers to a four-layer (business, service, network, and element management layers) object-oriented model to support telecommunication network management activities (details can be found in ITU-T's Study Group 4 activities at www.itu.int/ITU-T/studygroups/com04/index.asp, 2001).

**Toll Free** 1-800, 1-888, 1-877, 1-866, and other calling services for which the called party rather than the calling party is billed.

**TOS** Type of service; this refers to an 8-Bit field in the IP (both IPv4 and IPv6) packet header which indicates service priority (DiffServ Code Point or DSCP in a DiffServ domain) of the packet during enqueueing and emission from a queue.

**Transcoding** This refers to converting or transforming a previously encoded or compressed audio or video signal from one format (or compression scheme) to another. For example, transcending would be required if G.711 encoded voice signal from a POTS phone is delivered to an IP phone which can support only G.729 voice coding option.

**TRIP** Telephony routing over IP; an IETF-recommended (RFC 2871) telephony routing protocol that can work over any signaling protocol and is used for maintaining BGP-based routing information of IP-PSTN MGWs between different service providers.

**TTL** Time to live; a field in an IP header that indicates how long a packet is allowed to traverse a network as a valid entity before being dropped.

**UDP** User datagram protocol; a transaction-oriented IETF protocol (RFC 768) that can be used for low-overhead or unreliable communication in a packet-switched environment.

**UM** Unified messaging; this refers to a server based system which can store and process messages of any type (voice, text or e-mail, fax, etc.) of media via any type (telephone, computer, etc.) of interface.

**UPS** Uninterrupted power supply; this refers to a battery-pack and other protective circuitry based device to shield computers and data communication devices from power failure and related (like surge in voltage, current, frequency, etc.) damages.

**URI** Uniform resource identifier, a string of characters to identify an abstract or physical resource (see IETF's RFC 2396 for further details).

**URL** Uniform resource locator; this refers to an IETF specification (see e.g., RFC 1738, RFC 2732, and RFC 2806) which defines the syntax and semantics for representing the location of a resource (e.g., a file) in the Internet using a string of letters, numbers, and special characters. For example, the URL for IETF is http://www.ietf.org ("http" is the protocol to be used for retrieving the information, and "www.ietf.org" is the domain name for the resource or Web-site).

**USNO** United States Naval Observatory; this refers to an observatory that uses GPS to provide the official standard time within the United States (details can be found at www.usno.navy.mil).

**VLAN** Virtual local area network; this refers to services and applications based logical grouping of LAN terminals or workstations irrespective of the physical location of the workstations in the LAN segment. The objective is to achieve load balancing and bandwidth allocation for critical applications

—like VoIP—when both PCs and IP phones share the same network (see www.ieee802.org/1/pages/802.1Q.html, for further details).

**VoiceXML**   An incarnation of XML, which can be used for controlling and managing the call flow in an IVR system, listening to the content of a Web page via phones, etc. (details can be found at www.voicexml.org, and w3c.org/voice).

**VoIP**   Voice over IP; this refers to transmission of real-time telephone quality speech or voice signal—after digitization and packetization—over an Internet protocol (IP) based network—an Intranet or a VPN over the Internet—with or without sacrificing POTS-like reliability, quality, and availability.

**VoIP Call Server**   This refers to an IP based network element which controls setup of VoIP calls, and directly or indirectly manages IP-PSTN media gateways (see also the definitions of MGC and CC).

**VoMPLS**   Voice over MPLS; carrying real-time voice packets openly over MPLS without encapsulating them using IP, as described in the implementation agreement that has been developed in cooperation with ITU-T study groups 11 and 13 (SG 11 deals with signaling requirements and protocols issues, and SG 13 deals with multiprotocol and IP-based networks).

**VPN**   Virtual private network; a network that has been implemented by overlaying point-to-point links over leased lines or over the public Internet. This can be implemented in the networking hardware/firmware using a software-only solution, such as Microsoft's point-to-point tunneling protocol (PPTP). When the public Internet is used, all communication must be encrypted to ensure security.

**VRU**   Voice response unit; a personal computer or server-based system, which can accept incoming telephone calls, and can respond to customers' queries (entered via DTMF-/touch-tone or voice phrase) by playing voice files, updating customers data files, transferring the call, making outgoing calls, and so on. VRU and IVR are sometimes used synonymously.

**WAN**   Wide area network, see LAN/MAN/WAN for details.

**WFQ**   Weighted fair queueing; a queueing algorithm that allows enqueueing of the incoming packets as per pre-specified weights of the service within the desired fairness constraints.

**WG**   Work group; a focused group within any organization, for example, IETF has multiple WGs with each broad area—like Internet, routing, security, transport, etc.—of activity.

**WLL**   Wireless local loop; this refers to radio-frequency (like LMDS and MMDS, as discussed in Chapter 7) and free-space-optical signal based systems that connects the customers directly to the publics switched telephone network (PSTN), substituting the twisted pair of copper wires of the wireline local loop.

**XML**   Extensible markup language; a schema (arbitrarily defined vocabulary)-based standard format for defining structured documents and data on the Web (details can be found at w3c.org/xml).

# INDEX

CL

621.
385
KHA

6000653231